SOLAR FLARES AND COLLISIONS BETWEEN CURRENT-CARRYING LOOPS

SOLAR FLARES AND COLLISIONS BETWEEN CURRENT-CARRYING LOOPS

TYPES AND MECHANISMS OF SOLAR FLARES AND CORONAL LOOP HEATING

Edited by

JUN-ICHI SAKAI

Laboratory for Plasma Astrophysics and Fusion Science,
Toyama University, Japan

and

CORNELIS DE JAGER

SRON Laboratory for Space Research,
Utrecht, The Netherlands

Reprinted from *Space Science Reviews*, Vol. 77, Nos. 1–2, 1996.

KLUWER ACADEMIC PUBLISHERS

DORDRECHT / BOSTON / LONDON

A C.I.P. Catalogue record for this book is available from the Library of Congress

ISBN-13: 978-94-010-6635-8 e-ISBN-13:978-94-009-0331-7
DOI: 10.1007/978-94-009-0331-7

Published by Kluwer Academic Publishers,
P.O. Box 17, 3300 AA Dordrecht, The Netherlands.

Kluwer Academic Publishers incorporates
the publishing programmes of
D. Reidel, Martinus Nijhoff, Dr W. Junk and MTP Press.

Sold and distributed in the U.S.A. and Canada
by Kluwer Academic Publishers,
101 Philip Drive, Norwell, MA 02061, U.S.A.

In all other countries, sold and distributed
by Kluwer Academic Publishers Group,
P.O. Box 322, 3300 AH Dordrecht, The Netherlands.

Printed on acid-free paper

PREFACE

In this volume we compare modern observations of solar flares with results from recent theoretical research and simulation studies on current-carrying loops and their interaction. These topics have undergone rapid developments in the course of recent years. Observational results by X-ray monitoring and imaging spacecraft in the seventies and by dedicated imaging instrumentation in the satellites *Solar Maximum Mission* and *Hinotori*, launched 1980 and 1981, have shown the importance of X-ray imaging for understanding the ignition processes of solar flares. Such observations, in turn, stimulated theoretical studies, centered around the flux-tube concept. The classical idea that flares originate by interaction of current-carrying loops was developed and proved to be promising. Concepts on reconnection and coalescence of flux tubes were developed, and their consequences studied. The Yohkoh spacecraft, launched 1991, showed the overwhelming importance of coronal flux tubes and their many possible ways of interaction. Subsequent and parallel theoretical studies and simulations, differentiating between the topology of interacting fluxtubes, demonstrated that the mutual positioning and the way of interaction are important for the subsequent processes of energy release in flares and the many associated phenomena such as the expulsion of jets and the emission of X-ray and microwave radiation. The new developments now enable researchers to understand and classify flares in a physically significant way. Various processes of acceleration are active in and after flares on greatly varying timescales; these can now be distinguished and explained.

We are of the opinion that a comprehensive review of flare observations, theoretical studies and computer simulations is timely. That was our main motive for writing this book. We also strongly realize that our work is far from being an endpoint; it is rather a progress report: research continues and we hope that this review will help stimulating further research, observationally as well as theoretically.

Toyama — JUN-ICHI SAKAI
Utrecht — CORNELIS DE JAGER

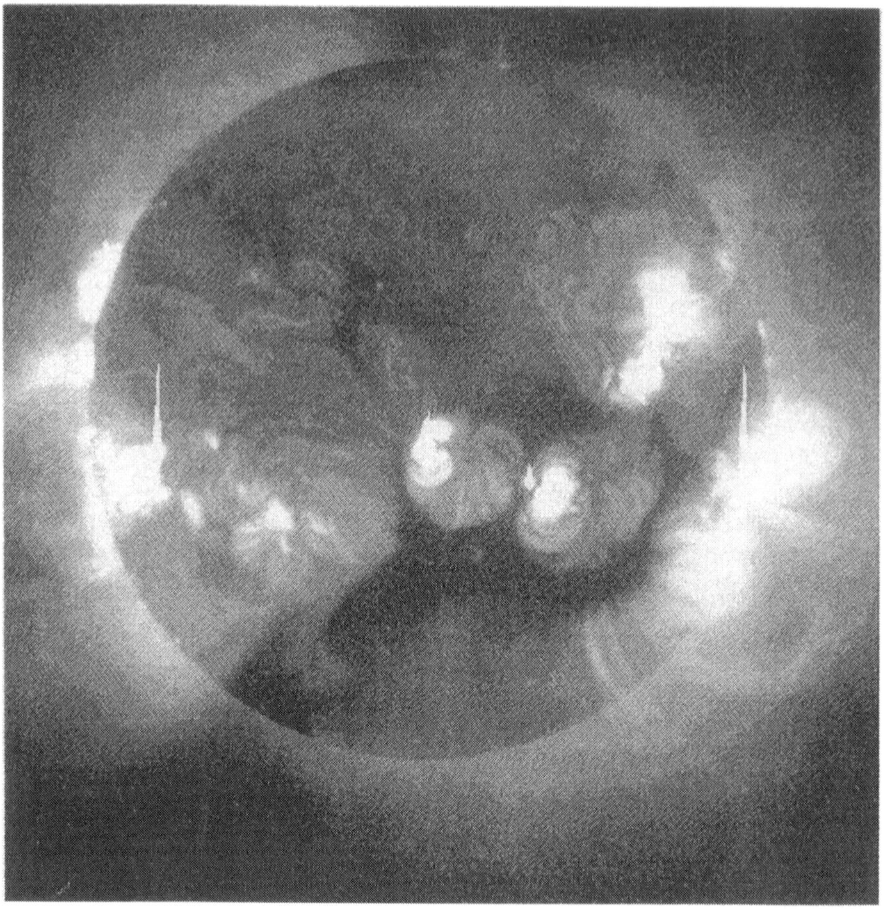

The Sun in X-rays observed by the Soft X-ray Telescope aboard
Yohkoh on 29 November 1991, at 21:11:45 UT; Al1-filter

SOLAR FLARES AND COLLISIONS BETWEEN CURRENT-CARRYING LOOPS

Types and Mechanisms of Solar Flares and Coronal Loop Heating

JUN-ICHI SAKAI

Laboratory for Plasma Astrophysics and Fusion Science, Department of Electronics and Information, Faculty of Engineering, Toyama University, Toyama 930, Japan

CORNELIS DE JAGER

SRON Laboratory for Space Research, Sorbonnelaan 2, 3584 CA Utrecht, The Netherlands

(Received 12 February, 1996)

Abstract. This paper deals with the temporal dynamics of solar flares. It gives a systematic description of solar flare models and tries to link the observations to results of simulations.

After a review of the development of ideas on flare structure and on theories on current-loop interaction in flares since the pioneering work by Gold and Hoyle (1960), this paper gives first a synthesis of present-days observationally based views on solar flares, essentially describing the developments since the review by de Jager (1986). We distinguish between confined/impulsive and eruptive/dynamic flares (briefly: confined and eruptive). The main difference between these two types is one of field-line topology: 'closed' or 'open'. The 'grand instability' in a field-line system opening to space is basic to the relation between eruptive flares, filament instability, and Coronal Mass Ejections.

A fair part of the paper deals with the developments in our understanding of the physical processes during collisions between current-carrying loops. After work by Tajima *et al.* (1982), who introduced the concept of current-loop coalescence in solar flares, using results obtained from two-dimensional particle simulation, it became clear that the current-loop interaction process includes a rich variety of physical mechanisms associated with rapid magnetic energy conversion through partial or complete magnetic reconnection, such as prompt high-energy particle acceleration, plasma heating, shock formation, plasma-jet formation and plasma radiation. This part of the paper concentrates on the developments since the review by Sakai and Ohsawa (1987), dealing with particle acceleration by magnetic reconnection and shocks during current-loop coalescence in solar flares. Theoretical research performed since the above review paper refers to magnetic reconnection, shock formation, particle acceleration and plasma emission during collisions between current-carrying loops.

These theoretical developments are compared with observations. The *Yohkoh* spacecraft, launched August 30, 1991, observed many evidences of two-loops interaction in active regions and a growing number of examples of flares due to coalescence of loops. These observational evidences are reviewed in a systematic way (Section 6), by relating them to the theoretical studies described before. First we describe flares due to interactions of the Y-, X-, and I-types. These can be either confined or eruptive. A new observation is the occurrence of 'ongoing partial reconnection' in a flare. Thereafter, we discuss flares due to single-loop instability (these are confined flares), followed by the newly discovered 'cusp-type' flares. These are eruptive flares. We end with a description of a typical eruptive flare. In Section 7 we summarize observational and theoretical evidences for our earlier proposed scheme of three different phases of acceleration in flares. The summary in the last section compares flare observations with flare simulations; it deals briefly with the three phases of acceleration, and outlines the present status and prospects of simulation studies and related theoretical research.

Space Science Reviews **77:** 1–192, 1996.

Table of Contents

1. Introduction

1.1. Some History

Magnetic fields play an important role in the solar atmosphere. Particularly in the upper atmosphere, the chromosphere and corona, the fields determine the dynamical features. On the Sun, the magnetic field is mostly concentrated in magnetic flux-tubes. These are the main channels for the transport of energy, generated in lower photospheric levels, into the corona. One of the manifestations of such a magnetic structure is the coronal loop. The loops are located in active regions and their footpoints are anchored near sunspots with opposite magnetic field polarities.

Longitudinal currents are excited in the magnetic loops due to the complicated motion of sunspots and the surrounding photosphere. In the earlier literature (Sheeley *et al.*, 1975; Švestka and Howard, 1981; Howard and Švestka, 1977) several possible examples of reconnection of magnetic field lines, taken from *Skylab* observations are presented. The reconnections give rise to sudden enhancements of the loop emission observed in soft X-rays. During the last few decades, interacting coronal loops have been investigated in relation to flares (Machado, 1982) and it has been shown that most confined/impulsive X-ray flares have their origin in two or more interacting loop structures. Also, high-quality observations of post-flare coronal loop systems recorded in visible emission lines reveal transient enhancements due to localized loop interactions. Smartt *et al.* (1993), reviewing the morphology of these enhancements based on ninety events, have found that they show occasionally marked enhancements in the vicinity of the projected intersection of two loops. Šimberová *et al.* (1993) presented evidences of the interaction of large-scale loops high in the corona, from an analysis of *Skylab* X-ray pictures of a giant coronal arch.

The *Yohkoh* spacecraft, launched on August 30, 1991, gave new impetus to the observational study of interaction between loops, thanks to the high image resolution of the Soft X-ray telescope, being only 2.4 arc sec at 8 Å. The soft X-ray images (cf., Shimizu *et al.*, 1992: Figure 1) obtained by this satellite showed the existence of many coronal loops, with typical temperatures ranging from 3 to 8 MK (Kano and Tsuneta, 1996). Often, such loops are coupled in active regions of the solar atmosphere, which may give rise to transient X-ray brightening. A considerable number of evidences, showing two-loops coalescence in flares observed by *Yohkoh* have so far been found.

It was a long way, though, before the relationship between flares and interactions between loops was established and understood. Although the flare phenomenon is known already since the middle of the 19th century, and in spite of the establishment in 1934 of a global flare patrol service based on spectroheliograms, truly systematic studies could only be undertaken since about 1940, after the invention of the narrow-band Hα filter, independently by Lyot and Öhman. Flare cinematography showed the quasi-explosive nature of many of these structures. These studies,

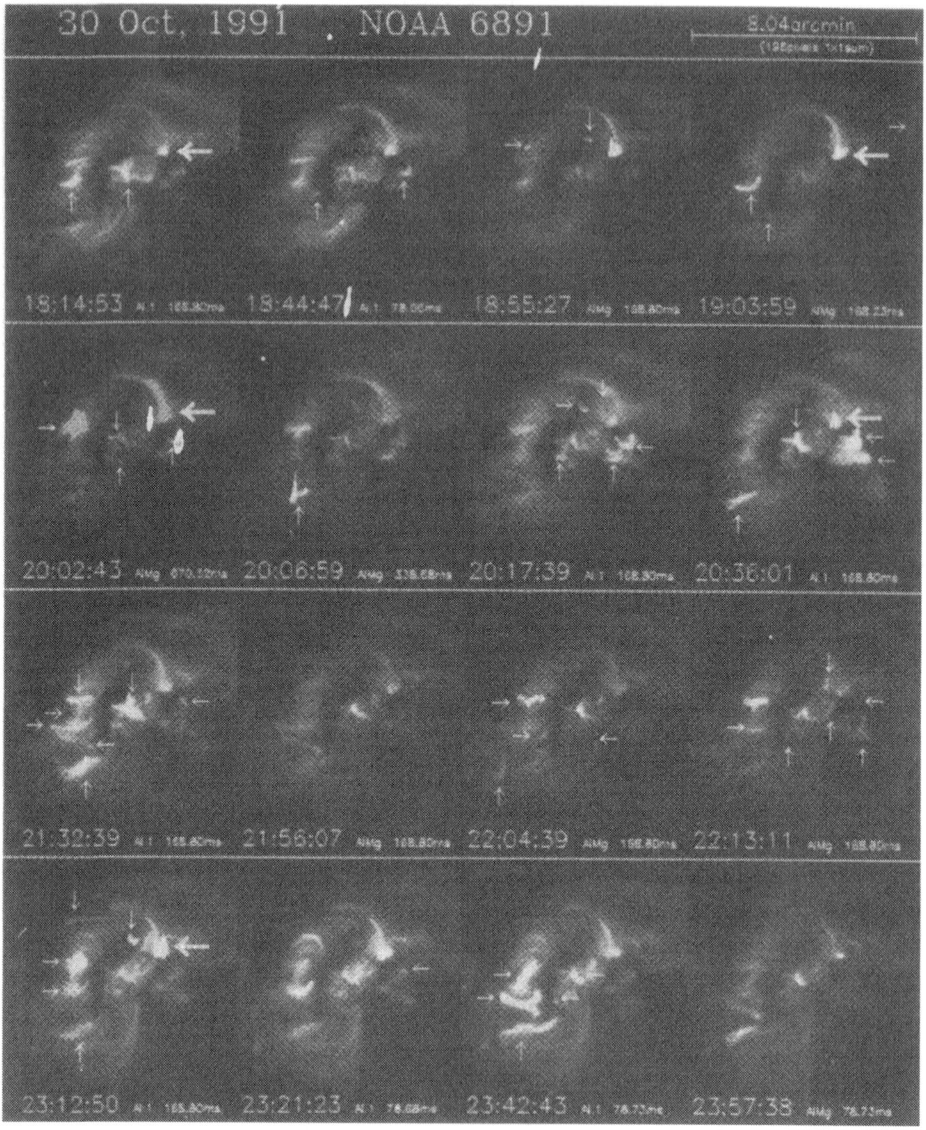

Figure 1. A sequence of soft X-ray images of a part of the Sun, showing many loops and their spatial and temporal developments. Arrows indicate locations of intense transient brightening. (Shimizu *et al.*,1992. From *Publ. Astron. Soc. Japan.*)

however, as well as spectrographic studies in the visual spectral range could only reveal the cool flare component (the 'Hα-flare'), which is mainly characterized by its sudden increase of the particle density, reaching values of the order of $n_e = 10^{13}\,\mathrm{cm}^{-3}$. Only in 1958 have microwave flare studies (Hachenberg, 1958) and X-ray observations from rockets (Friedman *et al.*, 1958) demonstrated the

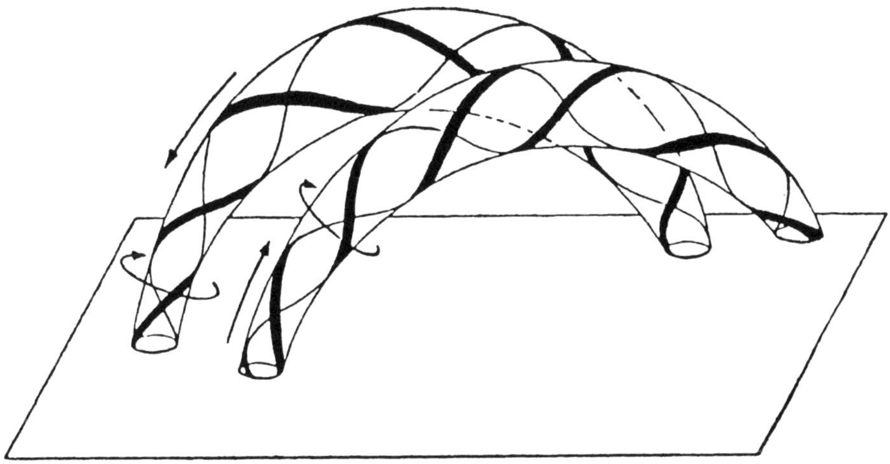

Figure 2. Gold and Hoyle's (1960) model for current-loops interaction leading to a flare. (From *Monthly Notices Roy. Astron. Soc.*)

existence of a *high-energy flare component* with estimated kinetic temperatures in the range of 10^6 K to 10^7 K. These observations raised the question of the source of the flare energy. Since most flares occur in magnetic regions close to sunspots, with preferences for the more complicated (δ-type) magnetic configurations, the source was sought in the conversion of magnetic to kinetic energy.

The idea that reconnection between magnetic fluxtubes may be important for explaining solar flares originates from that period. The pioneering work was done by Gold and Hoyle (1960), who elaborated on earlier work by Giovanelli (1946) and Sweet (1958). Gold and Hoyle proposed a model for current-loops interaction in solar flares (Figure 2). The suggestion was taken up and applied to solar as well as to magnetospheric problems by Dungey (1961), Parker (1963) and Petschek (1964). In his review paper, Sweet (1961) elaborated the reconnection model for solar flares and he showed the importance of the current-sheet, i.e., the region of interaction of the antiparallel magnetic fields of the two loops, where a steady electric current develops.

For a number of years the so-called Dungey–Sweet model and the Petschek model of flares were popular, in spite of the fact that quantitative checking did not yet appear possible. It is true that in the mean time the first satellites carrying soft X-ray recording instruments had been launched (the US SOLRAD-series; the UK *Ariel*, Europe's ESRO-2, and USSR spacecraft), but these showed only the 'gradual rise and fall' pattern, characteristic for flares seen in soft X-rays, of which we know now that it manifests mainly the flare's aftermath, not its impulsive ignition.

The process of flare instability came nearer to observational verification with *Skylab* imaging observations (1973). These were preceded by soft X-ray solar images obtained from rocket-borne X-ray telescopes. The new results showed the importance of loop structures in flare regions. The launch of satellites with hard

X-ray recording equipment (ESA's TD1A satellite) at about the same time (1972) showed that certain flares, those of the impulsive type, emit hard X-ray bursts during one or a few minutes in the early phase of a flare. These bursts occur as a rule during the rising phase of the soft X-ray intensity curve $I(t)$ and are mostly found at times near the inflection point of the $I(t)$-curve, where $\mathrm{d}^2 I/\mathrm{d}t^2 = 0$. The bursts occur in groups and were named 'Elementary Flare Bursts' (van Beek *et al.*, 1974; de Jager and de Jonge, 1978). In retrospect, the adjective 'elementary' was not well-chosen; there is a hierarchy of bursts (Kiplinger *et al.*, 1983) and we now propose to call them *Impulsive Phase Bursts*. It was hypothesized that these bursts were manifestation of the flare ignition process, by loop interaction.

These observational developments stimulated theoretical research on the process of interaction between loops. Sweet's ideas of the importance of the current-sheet were taken up and elaborated in Russia by Syrovatskii (1975) and his pupils, notably Somov, cf., Somov and Syrovatskii (1974, 1977, 1982), Somov and Titov (1985a, b), and Somov and Verneta (1988, 1989, 1993). We refer also to Somov's book (1992) which summarizes these and related studies.

The investigations described so far were all based on the MHD-approach and the idea of steady reconnection. New observational developments came in the early eighties when powerful solar satellites: the *Solar Maximum Mission* and *Hinotori* gave a wealth of information on impulsive phenomena in the early phases of flares, and showed the importance of loop-interaction for the flare process. It became clear that flare ignition in loops overlying the magnetic inversion line produces energetic particles that bombard the lower-lying chromospheric regions, causing footpoint evaporation (ablation): hot plasma filling the loops and spreading over the area: the soft X-ray burst, followed by the coronal explosion. This scenario is summarized in the notion *Impulsive Phase Explosion*. Impulsive gamma-ray emission became another new signature of impulsive flares.

These observations and the developments of new views on the process of current-loops interaction and its numerical-mathematical treatment were at the bases of a line of research that started when Tajima *et al.* (1982) presented their model of current-loop coalescence (essentially: interaction of current-carrying loops that are not necessarily anti-parallel in their magnetic field vectors) in solar flares and coronal brightening, by using results obtained from two-dimensional particle simulation. It became clear that the current-loops interaction process includes a rich variety of physical mechanisms associated with rapid magnetic energy conversion through explosive magnetic reconnection, such as prompt high-energy particle acceleration and plasma jet formation (Tajima and Sakai, 1989a, b), which is different from the case of steady reconnection (Sweet, 1958; Parker, 1963; Petschek, 1964).

1.2. Current-Loops Coalescence

The current-loops coalescence model (Tajima *et al.*, 1987) provided keys for understanding many of the characteristics of solar flares, such as explosive plasma heating, high-energy particle acceleration for both protons and electrons, and quasi-periodic oscillations of the electromagnetic emission. Tajima *et al.* (1987) tried to explain two solar flares (those of 7 June, 1980 and 26 November, 1982) in terms of the current-loops coalescence model. Sakai and Ohsawa (1987) presented a review paper on particle acceleration by magnetic reconnection and shocks during current-loops coalescence in solar flares. After the above review paper it became clear that the processes of loop-coalescence may have different signatures (Sakai, 1989, 1990; Sakai and de Jager, 1989a, b, 1991), depending on the geometry of the region containing the two interacting current-loops. The key parameters characterizing these different signatures are the scale L along the current-loop, which characterizes the length of the interaction region of the two loops, and the radius R of the current-loop, as shown in Figure 3. In the case shown in Figure 3(a), where$L \gg R$, an almost one-dimensional current-sheet can be induced in the interacting region by a quasi-parallel approach of the two loops. We call this type I-type coalescence. In this situation the current-loops coalesce with quasi-periodic oscillations when B_p (the magnetic field produced by the current) exceeds B_t (the magnetic field along the loop). According to Sakai (1990), electrons and protons can both be simultaneously accelerated to relativistic energies within one second during the process of coalescence. In contrast, when $B_p < B_t$, the loop-coalescence is only associated with motions causing plasma tilting without strong quasi-periodic oscillations.

When $L > R$, as in Figure 3(b), a flow of strong plasma-jets can be driven by the plasma tilting motion around the point of magnetic reconnection (Sakai, 1989). We call this Y-type coalescence. This mechanism has been applied to explain coronal explosions (Sakai and de Jager, 1989b).

The third situation is shown in Figure 3(c): $L \approx R$. This is the case of three-dimensional X-type current-loop coalescence (Sakai and de Jager, 1989a). This case, where the two crossed flux-tubes interact in one point, is a fundamentally new process as compared with the above two cases. In the I- and Y-type models the currents in the two loops are assumed parallel. If they were anti-parallel, there would be no attraction, but rather separation of the two current-loops, hence no reconnection. Little or no energy would be released.

These three types of loop-interaction were observed in soft X-ray images taken from *Yohkoh* (Shimizu *et al.*, 1992). A few flares, those of 17 August, 1992 (Takahashi *et al.*, 1995) and 9 December, 1992 (Inda-Koide *et al.*, 1995) showing I-type coalescence were observed by *Yohkoh*. A flare showing X-type coalescence was observed on 2 August, 1993 (de Jager *et al.*, 1995). Shimizu *et al.* (1994) found that 25% of the observed cases of two-loop interaction are of the X-type, 61% are of the Y-type (interaction close to the footpoints) and only 2% are of the I-type.

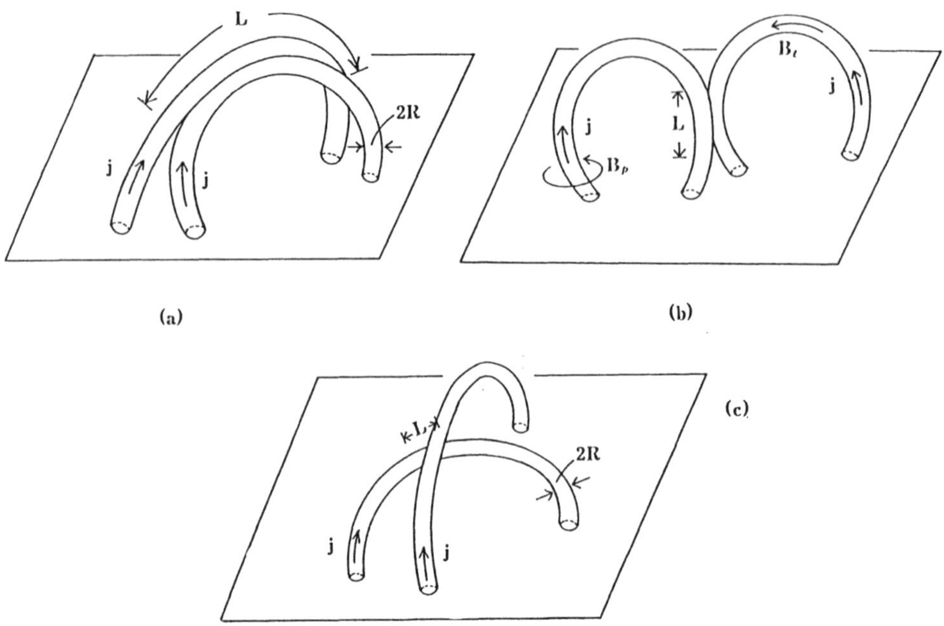

Figure 3. Schematic pictures showing three types of current-loops coalescence: (a) 1-D coalescence (I-type), (b) 2-D coalescence (Y-type), (c) 3-D X-type coalescence. L is the characteristic length of the interacting region. $2R$ is the diameter of the loop with plasma current **J** along the magnetic field B_t. B_p is the poloidal magnetic field produced by the plasma current. (Sakai and de Jager, 1991. From *Solar Phys.*)

The mechanism for the origin of flares on the basis of current-loops collision can only work if there is some kind of driver; there must already exist an initial instability, causing the approach of two or more fluxtubes, or – if one is dealing with a single fluxtube – the fluxtube must experience another driver leading to instability, such as footpoint rotation. We wish to emphasize that such kinds of instability are actually the essential properties of the pre-flaring region. To that end one has to realize that the sunspot areas are in regular, sometimes fierce, often shearing, motions and that those areas in which the shear is strongest and the field distribution most complicated, have the largest probability of producing flares (Kovacz and Dezsö, 1986; Gesztelyi and Kalman, 1986; Gaizauskas and Harvey, 1986, and references there to other publications). In adddition, the pre-flaring area is permeated by a number of fluxtubes that originate from small areas at the photospheric level in the active region. These areas, with characteristic sizes of about 10^4 km, have been discovered in hard X-ray images of flares and are usually called the 'footpoints' (Hoyng *et al.*, 1981; Machado *et al.*, 1982). These footpoints are the primary seats of the flares at the chromospheric/coronal level, as seen in hard X-rays.

The loops, connecting the footpoints, are not homogeneous. Evidence has been summarized by de Jager (1986a, p. 51). From X-ray and spectral studies (which give the electron density per unit volume and the total number of electrons along the line of sight) it is found that the loops as well as the plasma leaving the footpoints after the initial bombardment (the 'coronal explosion'; de Jager *et al.*, 1984; de Jager, 1985, 1986a, 1987) have an average filling factor of $\approx$ 0.01. This shows that a fairly substantial number of such thin loops, called fluxthreads, must emanate from the footpoints. These observations lead naturally to a model of the pre-flaring fluxtubes characterized by many fluxthreads. Estimates of the number of fluxthreads per loop range from ten to a hundred (cf., review by de Jager, 1986a). Theoretical simulations, described in Section 3.4 of this paper show that such filaments must develop if the footpoints of a current-carrying fluxtube are subjected to rotational motion. They are then wound up helically. They carry currents as well as counter-currents and are by their very nature unstable.

Apart from these observational and theoretical arguments supporting the helical fluxthread concept there is an indirect supporting consideration that can be derived from a statement by Klimchuk *et al.* (1992). The loops, observed with *Yohkoh*, are very much longer than their diameters and run through parts of the corona in which the pressure varies over many powers of ten. Yet, the loops have the same diameter, all over their length. That can only be so if the loops are filled for only a small fraction of their apparent volume with X-ray emitting fluxthreads, or – stated differently – if a loop is the optically unresolved complex of many thin (probably helical) fluxthreads.

The fluxthreads, being anchored in the photospheric level, undergo irregular mutual motions by the shearing and rotational motions in the photosphere, and are regularly moving with respect to each other. Additionally, emerging fluxtubes are often already sheared when rising from below (Leka *et al.*, 1996). Therefore, the pre-flaring area is never in a state of static equilibrium. At most times motion is a regular feature of the area, and it is these motions that are expected to give rise to interactions between fluxthreads of single loops or between the fluxthreads of two colliding loops. Conditions, that are favorable for the start of loop-collisions or of single-loop instabilities, are hence easily created at any time. Naturally, the stronger the shear, the larger the probability for a flare to occur, as has indeed been observed.

1.3. Outline of this Review

In this paper we review the present status of solar flare studies, and we emphasize and elaborate the difference between confined/impulsive and eruptive/dynamic flares (Section 2; part of that Section appeared earlier in a review paper by de Jager (1996), presented at the 1994 Sendai STP symposium).

In Sections 3–5 we review developments in the theory of physical processes during current-carrying loops collision, giving special attention to some applications to the mechanisms of solar flares and coronal loop heating. Research before

1987, related to current-carrying loops coalescence, was reviewed by Sakai and Ohsawa (1987).

In Section 3 we deal with the problems related to the generation of current-loops. It is shown there that a current-loop with a – sometimes fairly complicated – shell-shaped current structure can be generated due to rotational motion of the footpoints of the magnetic fluxtube. The loop, with its shell-current structure, becomes unstable against tearing instability when the intensity of the current-density exceeds a critical value. The nonlinear stage of the tearing instability, where many current-loops can interact with each other, is discussed, with applications to coronal loop heating and to solar flares.

In Section 4 we discuss the physical processes during coalescence of two current-loops. We show that there are two types of magnetic reconnection during current-loop coalescence: *partial reconnection* and *complete reconnection*. We review results of 3-D MHD simulations which show that during coalescence of two current-loops plasma jets may appear, sometimes with helical flow structures, and often associated with plasmoids. We also classify the processes of two-loops coalescence into the earlier introduced three types, the I-, Y-, and X-types, depending on their geometrical configuration. It is shown that these three types of coalescence have different physical characteristics, which opens the road to a comparison with the observations. Magnetic reconnection during two-loops coalescence is reviewed with special attention to high-energy particle acceleration. It is also shown that during coalescence of two current-loops fast magnetosonic shock waves as well as fast magnetosonic waves with finite amplitude can be generated and that these are important for understanding the mechanisms of high-energy particle acceleration and coronal heating.

In Section 5 we present results of kinetic effects associated with two-loops coalescence, that were obtained from 3-dimensional particle simulations. First we show the results of single-loop heating by a Buneman instability and by a kinetic kink-instability. Next we discuss the coalescence dynamics of two kink-unstable loops. We also describe the physical processes of high-energy electron acceleration during coalescence of two current-loops with partial reconnection, resulting from an analysis of a three-dimensional particle code. We discuss the electromagnetic wave emission process during two-loops coalescence.

In Section 6 we compare flare observations with the results of theoretical studies. We review observations of interactions between two loops, giving special attention to solar flares observed by *Yohkoh*, which are ordered according to the subdivision confined vs eruptive, as well as to subdivision between the X-, Y-, and I-types of two-loops interactions. One section deals with single-loop flares. Special attention is paid to the cusp-type flares, a discovery by *Yohkoh*, that may constitute a transition between the impulsive/confined and eruptive/dynamic flares; also for that reason it seems to be an important phenomenon for understanding the eruptive/dynamic flares. We describe a well-studied eruptive/dynamic flare.

Section 7 summarizes our views on the three phases of acceleration that may occur in fully developed eruptive/dynamic flares.

In Section 8 we summarize our views on the interpretation of flare observations, on the classification of flares and the scenarios of confined/impulsive and eruptive/dynamic flares; on the various phases of the acceleration processes, and on the status and prospects of simulation studies of the process of coalescence.

2. Flare Structure, Classification, and Models

2.1. Confined/Impulsive and Eruptive/Dynamic Flares

Helen Dodson Prince, well-known flare observer, with an impressive capacity of memorizing observations and their details used to say: *'Flares are different'*, and so it is. No flare is identical to another and the diversity is overwhelming if not embarrassing. That is the main reason for the failure of many a sophisticated attempt to establish ingenious systems for classifying flares. Wisdom, acquired over the years, has brought solar physicists to the viewpoint that complexity in flare classification does not necessarily add to clarity and understanding.

When, in the early eighties, after the launchings of the *Solar Maximum Mission* and *Hinotori*, abundant data became available on the high-energy aspects of flares, one learned to distinguish between the impulsive and gradual phases of flares. That stimulated solar physicists to call those flares that show these *two* aspects the *impulsive* flares. It also became clear that certain flares only show the gradual phase.

Thus emerged the simple subdivision of flares in *impulsive* and *gradual* flares (Pallavicini *et al.*, 1977). This classification is often used. At the end of the SMM/*Hinotori* period, after the drastic increase of the amount of information on energetic radiation and particles associated with solar flares, Bai and Sturrock (1989) presented a new subdivision of flares, viz.:

- thermal hard X-ray flares;
- non-thermal hard X-ray flares;
- impulsive gamma-ray/proton flares;
- gradual gamma-ray/proton flares;
- quiescent filament-eruption flares.

This scheme, however, has only little been used by others.

Subsequent studies, and particularly the data provided by the *Yohkoh* observations, with impressive information on loops and loop-interaction, stimulated looking for a subdivision, close to the impulsive vs gradual one, but adapted to the new observations. The new classification arranges flares in *confined* or *impulsive* and *eruptive* or *dynamic flares* (Švestka, 1985, 1995; Cliver, 1995). There is some correlation with the impulsive vs gradual classification in that sense that confined flares generally have a well-developed impulsive phase and that as a rule the eruptive flares last longer than confined ones, but the correlation is not a strict one. In this paper we will talk of *confined/impulsive* and *eruptive/dynamic flares*, although we will often call the former just confined and the others eruptive, for brevity. To some flares this division seems to offer problems: The *cusp-type flares* (Section 6.7) are fairly confined flares, but they have all characteristics of the eruptive/dynamic flares. The related *single loop flares* (Section 6.6) belong definitely to the class of confined flares. In addition, cases have been observed in which a confined flare triggers another, which may then lead to the development of an eruptive/dynamic flare. This shows the intricacies of this seemingly simple classification.

We turn in the following to a more detailed description of the two types.

2.1.1. *Confined/Impulsive Flares; an Overview*

Confined/impulsive flares are spatially compact and as a fairly general rule they show impulsive hard X-ray and/or microwave bursts. Such bursts nearly always occur in the early phases of flares. The question arises what the word ‘impulsive’ means. Zirin and Tang (1991), on the basis of observations with the Hard X-Ray Burst Spectrometer (HXRBS) aboard NASA’s *Solar Maximum Mission* (SMM), define impulsive flares as those where the half-power duration of the main hard X-ray spike is shorter than 30 s (for those events for which the all-channell HXRBS flux remains below 1000 counts s^{-1}), while events with a total countrate above 1000 s^{-1} are only called impulsive when the steepest rise of the main spike exceeds 500 s^{-2}. This nice attempt to quantify things has hardly been followed by others.

In the *optical range*, mainly in Hα observations, impulsive flares also show an increase in brightness in the early phase of the flares, but the many prominent spikes that are so characteristic for the microwave and hard X-ray spectral regions are as a rule not observed in Hα; the line width and intensity vary rather smoothly.

Further properties, elaborating the definition and the physical understanding of flares, depend mainly on the topology of the magnetic and velocity fields. All confined flares occur close to sunspots, and by preference in the most complicated (δ-type) magnetic fields. They are, fairly generally, situated over the magnetic inversion line, (i.e., the line where the longitudinal **B**-component is zero), and preferably in regions with a steep gradient of the longitudinal magnetic field component. The loops contain strong currents ($\simeq 10^{12}$ A) flowing through the current-loops. The largest currents amount to 30 mA m^{-2} (Gary and Démoulin, 1995). Flares occur at, or close to the locations where the currents are highest (Wu *et al.*, 1993a). These flares are spatially compact, hence the classification *confined*. They favour locations with high *magnetic shear* and strong transverse fields (Hagyard, 1992; Chen *et al.*, 1994). Zirin and Wang (1993) describe a flaring region showing a series of oppositely directed vertical field inversions, separated by thin layers of horizontal fields. It is generally thought that flares are due to conversion of magnetic energy into heat. However, while Sakurai *et al.* (1992) found clear evidence of relaxation of magnetic shear in the course of flares, Wang (1992) and Chen *et al.* (1994) describe cases in which shear appears to have increased after the flare. This may fit with the observation that only a very small fraction of the energy stored in a flaring active region is used for the flare.

Shearing *velocity fields* are important too. A flare on 15 July, 1992 was caused by interaction of two loops (Hanaoka, 1994). Some 24 hr before the flare the loops were only weakly sheared. Spot motion (0.25 km s^{-1}), that started thereafter, greatly increased the shear. The flare occurred at the intersection point of the sheared loops.

The *microflares*, mainly identified in the C IV 1548 Å emission line, with impulsive counterparts in soft X-rays (Porter *et al.*, 1995) also belong to the group of confined/impulsive flares.

2.1.2. *Eruptive/Dynamic Flares; an Overview*

The other group of flares, not obeying the definition 'confined', are spatially more extended and last longer. Often, their gradual phase last long, and it dominates the flare's Hα and soft X-ray lightcurves. For that reason they are often called *non-impulsive* or *gradual.* While this classification makes some sense for flares seen in the optical range, it does not seem to be well chosen in the energetic domain. In this paper we will show, with a number of examples, that the designation *eruptive/dynamic* fits better. One of the reasons for objecting to the designation 'gradual' is that many flares with a gradual intensity variation also have an impulsive start and could therefore also be called 'impulsive'.

The *eruptive/dynamic flares* can extend to some distance from sunspots, because, by comparison with the confined ones, they are larger. Most of them occur in well developed sunspot groups: the 'proton flares' are an example. Eruptive/dynamic flares reach greater heights than the confined ones, and are in many cases associated with filament eruption (Zirin and Tang, 1991), and with mass ejections from the corona. The statement that CME's (Coronal Mass Ejections) should bear no relation with flares (Gosling, 1993) is certainly exagerated (Švestka, 1995; Gosling and Hundhausen, 1995).

The microwave spectra of eruptive/dynamic flares peak at lower frequencies than those of confined flares.

2.1.3. *Different Classifications in Different Energy Ranges*

In relation to the above classification (that will be elaborated below) we have to note that there do exist differences of language in the definitions of flares between the three main groups of flare investigators: those concentrating on the *optical domain*, those studying the *hard X- and γ-ray* and/or *microwave* spectral regions, and those dealing with the *particle component.* In the optical range a flare is called impulsive if it is so in its Hα brightness or line-width variation, but that is seldom pronouncedly the case; to optical observations the classification *confined* definitely fits better. The confined flares tend to show bright Hα kernels in their early phases. Such kernels are also generally observed during the impulsive phases of flares seen in hard X-rays or microwaves. Finally, particle physicists call those

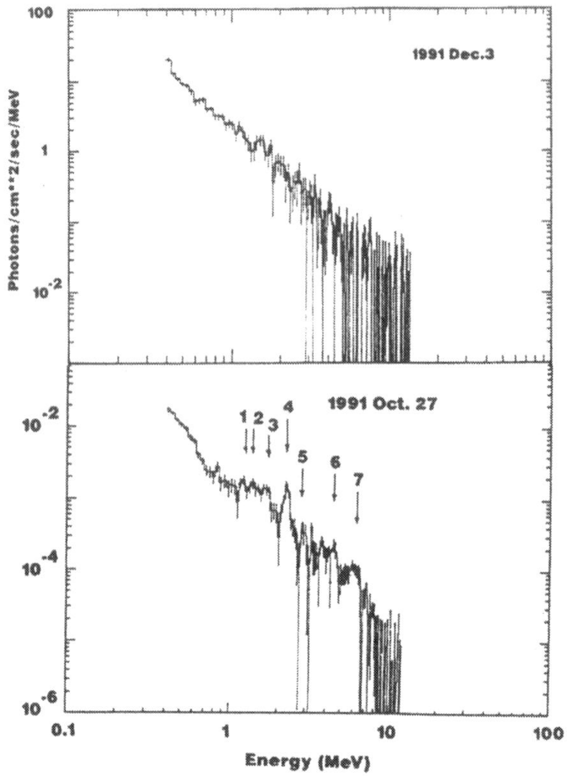

Figure 4. Upper diagram: gamma ray spectrum for the flare of 3 December 1991, showing mainly a continuous (Bremsstrahlung) spectrum. *Lower*: gamma ray spectrum for the flare of 27 October, 1991, showing lines of (1) Fe; (2) Mg; (3) Ne; (4) neutron capture; (5) O, Ne, and Mg; (6) C, and (7) O. (Yoshimori *et al.*, 1994. From *Astrophys. J.*)

flares 'impulsive' that show an impulsive onset of the particle flux, measured in interplanetary space. Obviously, this definition does not need to match with the former ones. Particularly, such flares need not show the impulsive hard X-ray signature.

In the particle domain one prefers to distinguish between two classes of flares (Ramaty *et al.*, 1993; see also Mandzhavidze and Ramaty, 1993): the first and the second class events. Those of the first class have a high electron-proton ratio (low for the 2nd class flares), are relatively rich in He^3 (poor in 2nd class flares), rich in heavy elements, have a high gamma-ray line flux, and have a low flux in solar energetic particles (SEP). The degree of Fe-ionization is high (+21) for first class events, lower (+14) for the others. Yoshimori (1994) found that the first class events correlate best with confined flares, the second class rather with the eruptive/dynamic ones. The eruptive flares correlate well with type II and type IV radiobursts, with coronal mass ejections (CME), and with interplanetary events having large Solar Energetic Particle (SEP) and proton fluxes (cf., section 2.4).

We note the importance of gamma ray observations (Figure 4) for diagnosing flares (Chupp, 1984, 1990; Yoshimori *et al.*, 1986). The *continuous gamma spectrum* has been observed in the energy range up to 10 MeV, with its main contribution below $\simeq$1 MeV. It is well described by a power law spectrum and it is emitted by electrons, chiefly those in the high-energy tail of the electron energy distribution curve. The *gamma-ray lines* are emitted in the spectral range between 1 and 7 MeV, and all but one are due to reactions between various kinds of nuclei with protons or α particles with energies up to a few tens of MeV's. There are two kinds of lines, prompt and delayed ones. The latter require the previous generation of other particles, such as neutrons, slowed down to the appropriate energy range. Gamma-ray observations are important for establishing the properties of the high-energy plasma in a flare. Its emission puts stringent demands and requires high densities. Slowing down and annihilation of positrons needs $\log n > 14$. For the decay of pions resulting from GeV protons, one needs $\log n > 14$, but the observation that neutrons can escape to space yields an upper limit to the density: $\log n < 15$. Positron annihilation demands $\log n_e \approx 16$. Hence, generally one may assume that the main part of the gamma-ray spectrum is formed in regions with densities $14 < \log n_e < 16$, which is fairly deep in the photosphere. For specific flares that range can be narrowed down considerably if suitable observations are available (Kawabata *et al.*, 1993; Yoshimori *et al.*, 1994).

2.1.4. *Main Properties of the Two Types of Flares*

We next list the *primary characteristics* of the two types of flares.

Confined/impulsive flares

– are compact,

– show low-lying loops,

– occur near to spots, preferably in complicated spot-groups,

– are situated over the magnetic inversion line, in a region of strong horizontal magnetic field gradient,

– have two or more *footpoints* situated on either side of the inversion line; these footpoints live as long as the hard X-ray bursts are lasting,

– are seats of current loops ($\simeq 10^{12}$ A),

– show the *Impulsive Phase Explosion* – cf., section 2.2,

– can have prompt gamma-ray emission, both in continuum and lines, with high corresponding temperatures,

– can have a high gamma-ray line flux,

– are *not* associated with radiobursts of types II and IV,

– can be associated with solar energetic particles (SEP), but are rarely associated with CME's; in case they are, the confined flare has triggered a eruptive/dynamic one which causes the CME (an example is the flare of 6 November, 1980),

– have an electron/proton number ratio that can vary greatly from one flare to the other,

– can have a number density ratio between He^3 and He^4 particles that is much larger than the 'cosmic' value, sometimes even approaching values of 0.1 to unity. There is also some relative enrichment of all light elements, like He^4 (up to a factor 10 with regard to H), Fe (up to a factor 7 to 10),and elements like Si, Mg, Ni (up to a factor 3).

Eruptive/dynamic flares

– occupy larger volumes,

– may therefore extend to larger distances from spots than confined ones,

– can extend higher into the solar corona than confined flares,

– are associated with filaments, and with changes in these, such as filament lifting, eruption, or *disparitions brusques* (sudden disappearances),

– last longer than confined flares,

– emit a microwave spectrum that peaks at lower frequencies than those of confined flares,

– have lower gamma-ray fluxes with lower associated temperatures than the confined flares,

– are associated with CME's,

– can have associated particle clouds with coronal abundances, and they do not show the He^3 enhancements,

– can have large SEP (proton) effects,

– are associated with type IV or/and type II radiobursts. The latter indicate that shock waves are moving through the corona with velocities of 1000 to 2000 km s^{-1}, while the moving type IV's are rather associated with a CME, and the stationary type IV's with a giant post-flare arch.

We do not wish to conclude this section on flare types without quoting from a letter by Dr Markus Aschwanden, kindly sent to us as a commentary to an earlier draft of this paper. His comments, in our view, illuminate well some of the problems and ambiguities met in defining flare types:

Hard X-ray observers are used to classify flares into 'impulsive' and 'gradual' types, which are almost synonymous to the 'confined/impulsive' and 'eruptive/dynamic' notions. The terms were born out of the morphology of the hard and soft X-ray time histories, and could therefore be applied without the knowledge of imaging data.

I fully understand the meaning of the 'eruptive/dynamic' flare type, if one attributes the descriptive adjectives to the MHD behaviour. Alternatively, one could easily argue that 'confined/impulsive flares' have also a dynamic and eruptive behaviour from the aspect of particle energization and transport. The adjectives 'impulsive' and 'dynamic' are synonymous words in general, rather than mutually exclusive. In this sense the terms 'closed' or 'open' magnetic field structures may serve better to characterize different dynamic aspects of the flare evolution.

Often, there is an apparently 'closed/confined' flare loop visible in soft X-rays, but the accelerated particles, perhaps originating from near the X-type reconnection point in the cusp above the flare loop, have apparently access to both con-

fined and open field lines, as detailed timing correlations between chromospheric X-ray spikes and coronal radio type III bursts demonstrate in many flares. Additionally, even in the presence of 'closed/confined' flare loops, 'eruptive/dynamic' phenomena such as soft X-ray ejecta are detected nowadays with deep-exposure Yohkoh-*maps with an increasing rate, fully consistent with the framework of the Kopp-Pneuman flare scenario. I don't want to stress the 'big flare syndrome', but the classification of flares into 'confined/impulsive' or 'eruptive/dynamic' types seems to depend often on what wavelength observers put their emphasis, while the distinction becomes more blurred with improved and more sensitive wavelength coverage.*

In this quotation Dr Aschwanden describes the 'cusp-type flares' treated by us in section 6.7. Like him, we consider these a transition between confined/impulsive and eruptive/dynamic flares.

2.2. THE IMPULSIVE PHASE EXPLOSION

2.2.1. *Three Phases in Impulsive Flares*

The spacecraft *Solar Maximum Mission* and *Hinotori*, launched 1980 and 1981, yielded essential new data on processes occurring in the early phases of flares. The data allowed researchers to elaborate the designation 'impulsive', a notion that should be described now. Rather than speaking of impulsive flares one should speak of flares with an impulsive phase. Out of such studies emerged the concept of the *Impulsive Phase Explosion.* This concept refers to the totality of the events observable during the impulsive phases of flares, and to their mutual causal relationships. Since most or all confined flares have an impulsive start, the Impulsive Phase Explosion is also part of confined flares. In this section we give a summary and a scenario for this phase, but we do not claim that every confined flare or every impulsive phase strictly follows that line of events. Since a fairly detailed commented description of the impulsive phase was given earlier by one of us (de Jager, 1986) we restrict this section to the essentials in so far as the 1986 description concerns; we add new information published since 1986 and that obtained by *Yohkoh* since 1991; we examine in how far this data modify and complete the earlier model.

A fully developed impulsive flare consists of three partly overlapping phases: the precursor, the impulsive and the gradual phases.

2.2.2. *The Precursor*

The *precursor* is a feature of which the existence had been suspected already since the late fifties. Bumba and Křivský (1959) were the first to study it and found that a preflare, as they called it, could occur within an hour before a flare and lasted on the average 5 to 10 minutes. One should, however, be cautious: the physical relationship between two flares following each other within one hour remains uncertain without detailed images. In Hα images the precursor is generally not very conspicious. From the later investigations in this spectral range we mention McCabe (1985).

Flare precursors in the radio-domain were described by Kai *et al.* (1983). It often happens that radio type III bursts which, as X-ray bursts, are assumed to be due to beams of accelerated electrons, may occur a few to ten minutes before the start of a hard X-ray burst (Benz *et al.*, 1983).

In the X-ray spectral region precursors were studied by Simnett and Harrison (1985) and Harrison *et al.* (1985). Schmahl *et al.* (1989) investigated the precursors in high-resolution X-ray images and found that they appear as loops or kernels, close to, but not neccessarily at the flare site. Tappin (1991) even concluded that most if not all X-ray flares have precursors, with properties not correlated with those of the main flares. A review was published by Simnett (1993). Many fine examples of precursors in the intensity-time curves of hard X-rays can be found in the *Yohkoh* HXT Databook (Kosugi, 1993).

Regular observations in the γ-region with SMM (Rieger, 1989) and *Hinotori* (Yoshimura *et al.*, 1986) enabled one to identify the gamma-precursor as a real constituent of flares. Chupp (1990) finds that many gamma-ray flares are preceded by precursors, in some of them even hours before the flare. He describes some precursors in a discussion of three flares observed with SMM. Apparently, the precursor is a well-defined element of the flare seen in very energetic radiations.

In spite of the fact that some bursts, identified as precursors, may eventually turn out to be chance coincidences, the above summary shows that in general precursors may be considered real parts of flares in most spectral regions studied. We proceed to a description of a few well-studied precursors.

The first observed spatially well-defined X-ray precursor appeared at least twice during a ten minutes period before the limb flare of 30 April, 1980 (the 'Queens's Flare') in an area with a size of a few thousand km, located about 10 000 km above the limb (de Jager *et al.*, 1983), close to the apexes of two adjacent loops (Figure 5). After these first brightenings the loop(s) as a whole became visible. Inda-Koide (1994) and de Jager *et al.* (1994, 1995) describe a few cases in which soft X-ray precursors were seen about ten minutes before impulsive flares.

Sakao *et al.* (1992) showed an instructive specimen of a precursor in the flare of 15 November, 1991 (Figure 6). The hard X-ray image was initially a single source, situated near the magnetic inversion line and it developed later into a double source in which the mutual distance increased with time (Figure 7). There were several spikes in the later part of the impulsive phase; these spikes were emitted from the footpoints. These observations suggest, as in the case of the Queens's Flare, that the precursor was emitted in the apex of a loop, and the obvious consequence is that this impulsive flare started with energetic particles being produced near the top of the fluxtube.

The feature shown by the precursors in Figures 5–7 is not common to all flares, though. From the study of a sample of 131 flares observed by *Yohkoh*, Farnik *et al.* (1996) found that 15% to 40% show precursors. The initial brightening occurs in

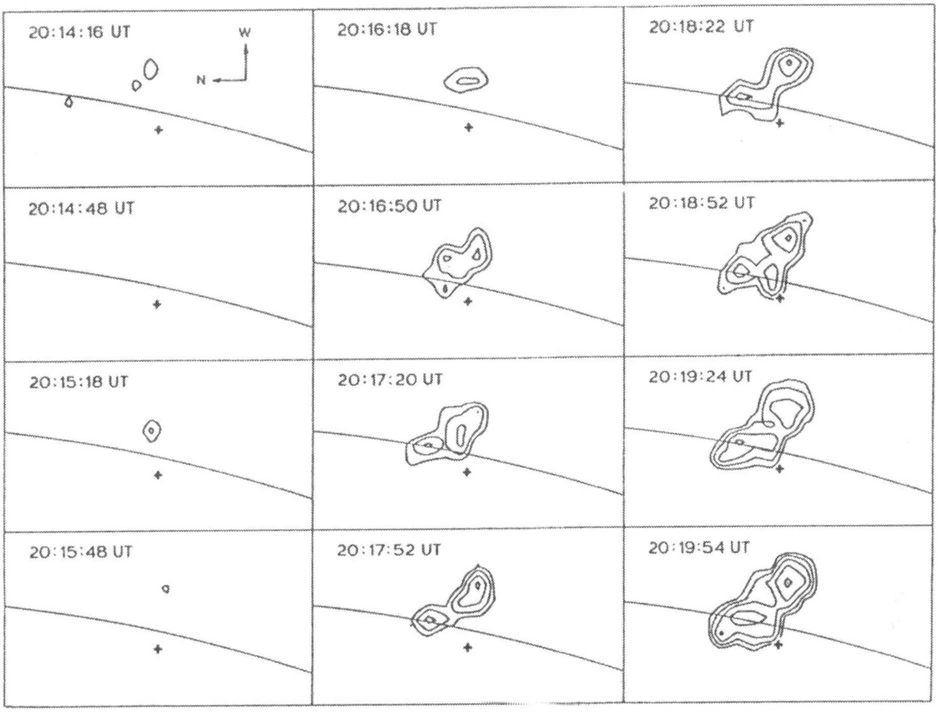

Figure 5. Soft X-ray images of the precursor of the limb flare of 30 April 1980, during a five minutes period.The flare started at 20:19 UT. (de Jager *et al.*, 1983. From *Solar Phys.*)

physically separate but nearby structures (Figure 8). Often, these structures share a footpoint with the structure that shows the flare brightening.

The precursor emission can be attributed to direct heating by the electric field in the coronal part of the loop (Dennis *et al.*, 1994). The subsequent bombardment of the lower chromosphere by energetic particles, originating near the apex, thereupon leads to the appearance of the footpoints. From these footpoints the impulsive hard X-ray bursts are emitted in a so called thick-target process. In other cases (such as the one shown in Figure 8) loop-interaction close to a footpoint seems to be the basic mechanism.

2.2.3. *The Impulsive Phase*

The *impulsive phase* proper is characterized by the emission of impulsive spike-like hard X-ray and/or microwave bursts. Neupert (1968) was one of the first to study the hard X-ray bursts. He discovered that they are strongest when the soft X-ray intensity-time curve reaches its inflection point: the time profile of the (gradual) soft X-ray emission matches the time integral profile of the hard X-ray bursts: the *Neupert effect*.

Subsequently, the hard X-ray bursts were studied by van Beek *et al.* (1974), and de Jager and de Jonge (1978) using observations from ESA's TD1A satellite.

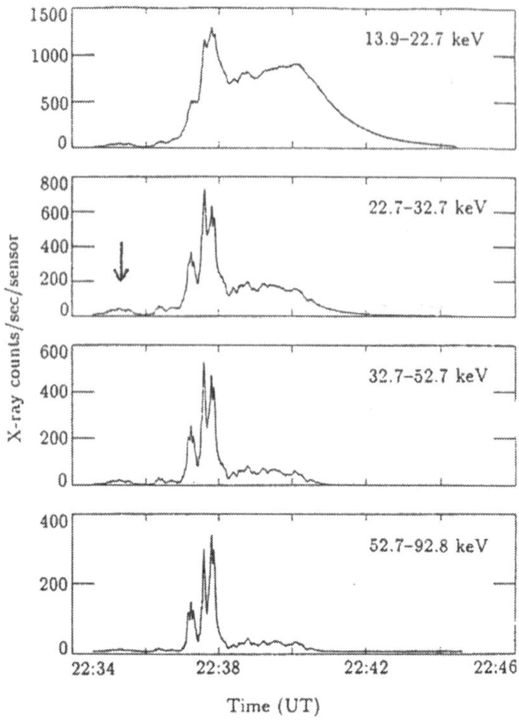

Figure 6. Hard X-ray time history of the flare of 15 November 1991. Note the precursor at 22:35 UT (Sakao *et al.*,1992. From *Publ. Astron. Soc. Japan.*)

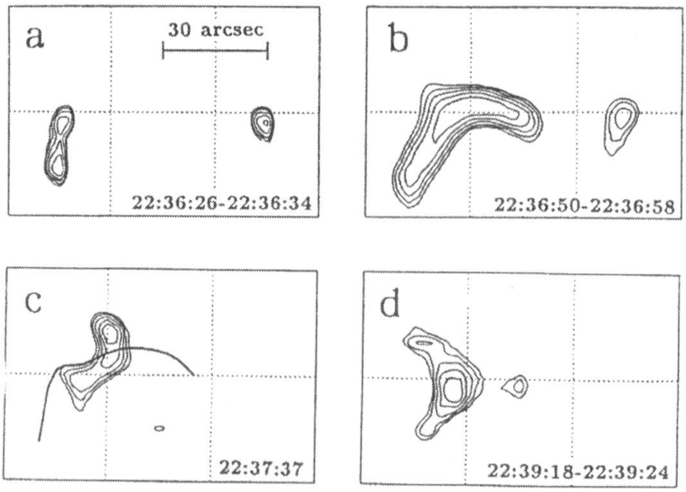

Figure 7. Evolution of the 14–23 keV X-ray source of the flare of 15 November, 1991, showing (a) the precursor, (b) just before the onset of the impulsive phase, (c) the impulsive and (d) the gradual phase. The line in image (c) is the magnetic inversion line. (Sakao *et al.*, 1992. From *Publ. Astron. Soc. Japan.*)

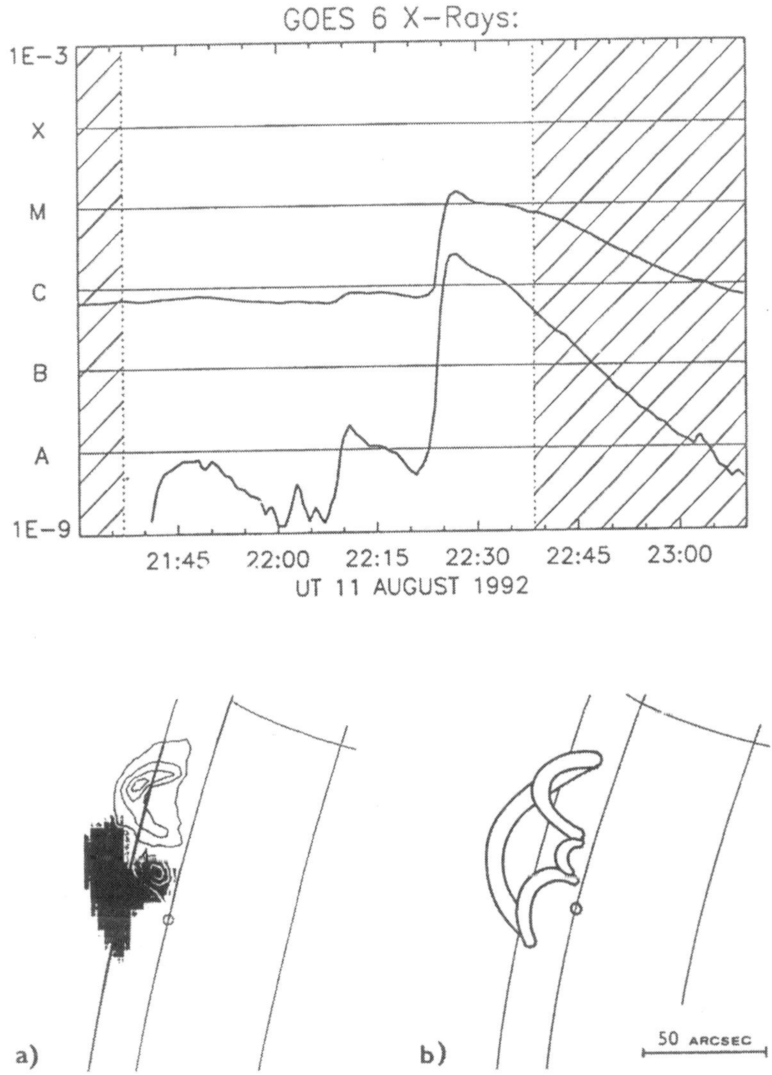

Figure 8. Precursor of the flare of 11 August, 1992. *Above*: GOES soft X-ray intensity-time curves. Shaded are the *Yohkoh* night intervals. *Below*: (a) Superposition of three *Yohkoh* soft X-ray images. *Black*: preflare emission at 21:48:15 UT; *white contours*: at 22:09:05 UT; *gray-scale images*: the flare emission at 22:34:19 UT. *Below*: (b) magnetic structure of the event. (Farnik *et al.*, 1996; from *Solar Phys.*)

They found that for one and the same flare the individual impulsive bursts are homologous in time: they have the same time-profile (Figure 9). This suggests that the basic emission conditions and mechanisms are the same for all impulsive bursts from the same flare. For another flare the time profiles are usually different. These

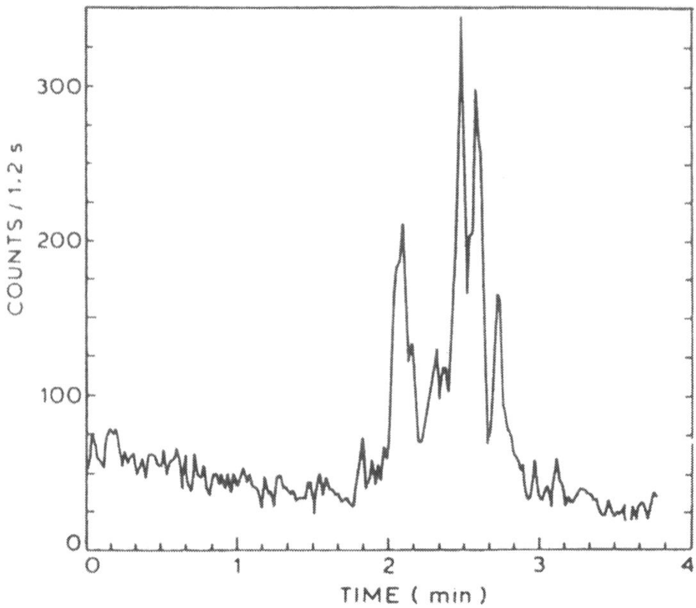

Figure 9. Impulsive Phase Bursts in the flare of 2 August, 1972. (de Jager and de Jonge, 1978. From *Solar Phys.*)

burst were initially called 'Elementary Flare Bursts'; we now prefer to call them *Impulsive Phase Bursts.*

The Impulsive Phase Bursts are related to the footpoints: small areas (10^4 km), normally situated at either side of the magnetic inversion line (Hoyng *et al.*, 1981; Duyveman *et al.*, 1982).

Canfield *et al.* (1992) noted the correlation between the locations of soft X-ray sources with sites of strong Hα red-wing intensity (Figure 10), which supports the idea that soft X-rays are emitted from the region where the Hα emitting sheet is moving downward under the influence of electron bombardment. They also found that hard X-rays are emitted from sites with a high vertical current density (Figure 11). This result was confirmed by Leka *et al.* (1993): the precipitation site of the non-thermal electrons is at the edge of vertical current channels.

Sakao *et al.* (1992), confirming earlier results, found that the double footpoint structure becomes particularly manifest at photon energies above 30 keV. The earlier SMM and *Hinotori* observations showed that below 15 keV the footpoints are hardly visible, and can only be made to appear by subtracting a general background. This is due (de Jager, 1986a) to the fact that electrons with energies below 15 keV have too small a mean free path to reach the deeper chromosphere with a sufficiently intense flux. Sakao *et al.* (1992) found clear evidence that the lower energy radiation as well as the radiation emitted in the absence of the impulsive bursts comes from areas in between the footpoints, in particular from the apex of

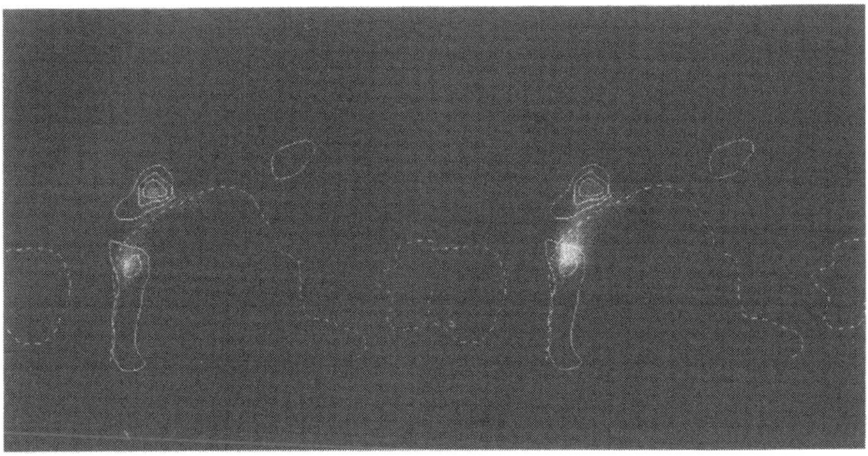

Figure 10. Soft X-ray images overlaid with Hα red-wing excess contours, before the start of the impulsive phase. There is correlation between the two. The dashed line is the neutral line. (Canfield *et al.*, 1992. From *Publ. Astron. Soc. Japan.*)

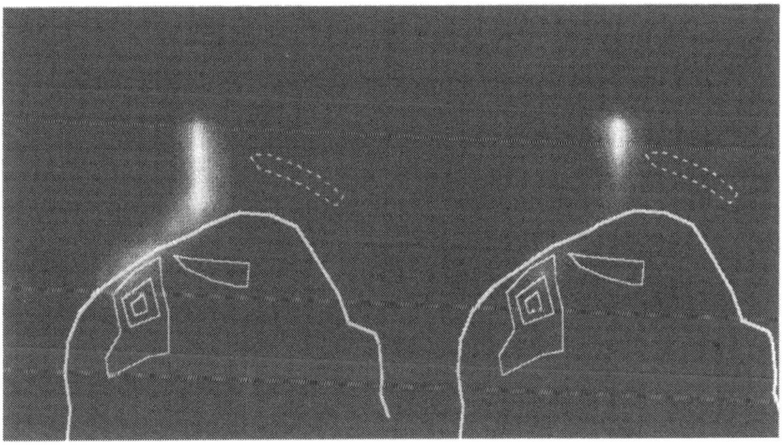

Figure 11. Hard X-ray emission. The vertical electric current-density is shown by the contours (positive solid, negative dashed). The white line is the neutral line. (Canfield *et al.*, 1992. From *Publ. Astron. Soc. Japan.*)

the loop connecting the footpoints, while the hard burst radiation is emitted by the footpoints (Figure 12). This discovery was only possible thanks to the high time and spatial resolution of *Yohkoh*. A case has been reported in which the four Impulsive Phase Bursts of one flare were emitted by four different footpoints (Costa *et al.*, 1994).

A fraction of 0.43 of the hard X-ray bursts have double structure and 0.28 have single (Sakao, 1993); the remaining ones are more complicated or cannot be defined. There are cases in which one footpoint is small with a strong magnetic field and the other is large with a weaker field. This is explained by the 'cornu

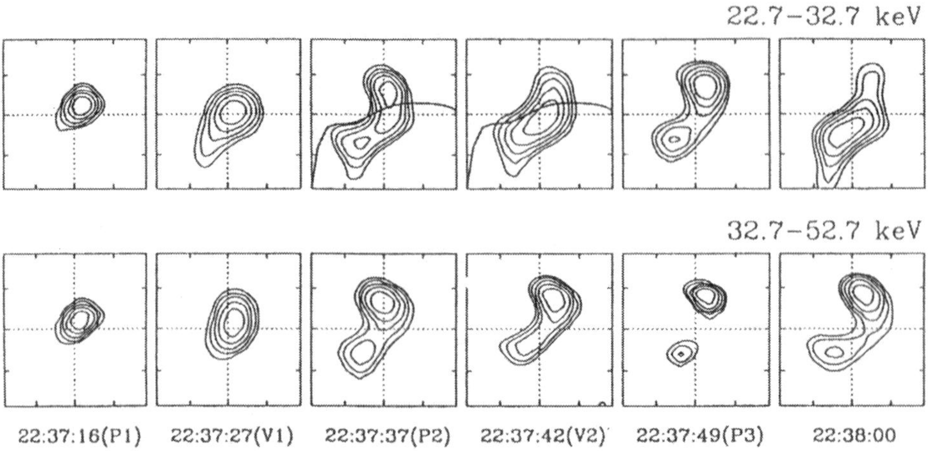

Figure 12. Hard X-ray images of the flare of 15 November, 1991 in two energy bands during the three impulsive spikes, showing that the hard impulsive phase bursts are emitted from the footpoints. P = peak and V = valley. (Sakao *et al.*, 1992. From *Publ. Astron. Soc. Japan.*)

copiae' model (Sakao, 1994): field lines converging towards the smaller footpoint and diverging towards the other.

2.2.4. *Footpoint Model*

The footpoints are attributed to the bombardment of the chromosphere by energetic particles from 'above'. We will show further in this section that these particles are electrons (but we mention a review paper by Simnett (1995) that describes the possible role of protons in the energy transport in a flare). How deep the electrons can penetrate, depends on their energy. Those with $E < 15$ keV do not reach the chromosphere (assuming usual loop densities), and those with MeV energies can penetrate into the photosphere and produce there the white-light footpoints (Hudson *et al.*, 1992). Such 'impulsive white light patches' (in contrast to 'gradual white light brightenings') usually exist for a few tens of seconds only, and are smaller than 15 arc sec (Sakurai, 1992b). Their sources are situated fairly deep inside the photosphere (Neidig *et al.*, 1993). In the photosphere, a temperature increase of 300 K has been reported for the white-light flare of 24 October, 1991 (Fang *et al.*, 1995). A sufficiently intense flux of very energetic electrons (MeV-range) is needed for producing them. A significant piece of information is that at least one of these white light footpoints was accompanied by the 511 keV annihilation line emission, that persisted longer than the electron bremsstrahlung and γ-ray events. This is attributed to positrons formed in the photosphere, that are partly diffused upward (Kawabata *et al.*, 1994). The positrons are chiefly due to deexcitation of excited O^{16} by e^-, e^+ pair emission. They have energies around 2.5 MeV. Other positrons, due to decay of nuclei like S^{31}, P^{29} have energies in the range of 1 MeV. It is assumed that half of them penetrate down to the photospheric layers with

$n_e = 2\times10^{16}$ cm^{-3} (for 2.5 MeV particles) or 10^{16} (for 1 MeV), coresponding to large optical depths ($>$ 10, hence subphotospheric layers).

The model described above can also explain how it happens that in two successive homologous flares the second flare did not show footpoints, as was the case on 31 August, 1980, when two about equally strong flares occurred at the same location briefly after each other. The second did not produce footpoints because the fluxtube was still filled with plasma from the first flare, which reduced the mean free path of electrons accelerated in the second flare: these could not reach deeper layers.

2.2.5. *Evaporation; Coronal Explosion*

Wuelser *et al.* (1992) find that the footpoint observations can be represented best by thick-target electron bombardment in an area of $(2.2 \pm 0.7) \times 10^7$ km^2, which would correspond with a circular area of diameter 5000 km. In these footpoints chromospheric gas is heated and it rises convectively upward with velocities of the order 150 to 400 km s^{-1}. These velocities have been observed spectroscopically (Antonucci *et al.*, 1984, 1985, 1990). From *Yohkoh* data, Culhane *et al.* (1992) find about 250 km s^{-1} and Doschek *et al.* (1992) report line-of-sight velocities of 200 to 350 km s^{-1}, from the flare onset onwards. They find a static component smaller than 50 km s^{-1}, but the lines show large Doppler broadening, up to the surprisingly large value of 800 km s^{-1}. Sometimes, larger upward velocities are observed, up to 800 km s^{-1} (Antonucci *et al.*, 1990).

The upward motion is not homogeneous but occurs in jets with sizes smaller than those of the footpoints (Borman and Lemen, 1992), which confirms the small filling factors found earlier. The filling factors, applying to the greater part of the flare, are of the order 0.01 (de Jager, 1986a; de Jager *et al.*, 1994). The inhomogeneity of the burst pattern finds support in high spatial and time resolution observations of impulsive phase bursts (Correia *et al.*, 1995). They find that bursts with durations longer than one second are superimposed bursts with components of subsecond duration. The fast pulses are emitted from sources separated by some 5 arc sec. The temperature of the rising gas is of order 30 to 50 MK (Tanaka *et al.*, 1982; Tsuneta *et al.*, 1984; de Jager, 1985b). The convection model is supported by the observation that the upward velocities are largest when the electron temperature T_{el} of the footpoints is highest in the soft X-ray rising part of the flare, while the upflow velocity decreases when T_{el} decreases (Culhane *et al.*, 1992). A thermal transition region with a typical temperature of the order 10^4 K (de Jager, 1987) originates between the heated gas and the undisturbed chromosphere, hence at the bottom of the evaporation 'pit' from where the heated gas emanates. Subordinate and resonance lines like Hα and Lα are emitted from this transition layer. In this thin layer ($\simeq$20 km thick; de Jager, 1985b) the particles undergo downward motion because of momentum conservation (Canfield *et al.*, 1987). This motion is observable by Hα redshifts (Ellison, 1949; Švestka, 1951; Ichimoto and Kurokawa, 1984; Canfield *et al.*, 1987); cf., also Figure 10. The location of the sources of these

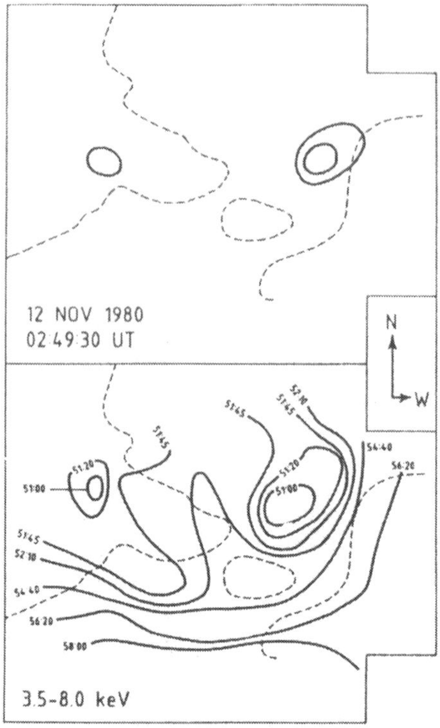

Figure 13. A coronal explosion in the flare of 12 November, 1980. *Upper diagram*: footpoints. *Lower*: isochrones of flare expansion; the lines are labeled with the time in minutes and seconds. (de Jager, 1985a. From *Solar Phys.*)

redshifts coincides with the footpoints seen in soft X-rays, while the hard X-ray bursts are emitted at the *edges* of the sites of high vertical component current density as shown in Figure 11 (Canfield *et al.*, 1992).

This process of ‘evaporation’ or ‘ablation’ also leads to plasma streaming out over the surrounding chromosphere, guided by the magnetic field line system. These ‘coronal explosions’ have been observed in soft X-ray images in the later phases of solar flares: plasma spreading with velocities of about 100 to more than 600 km s^{-1} (de Jager *et al.*, 1984; de Jager, 1985b, 1986a, b, 1987). Application of the conservation equations to the observations shows that the outflowing plasma moves in tubes with constant cross-sections and that the horizontal velocity component always decreases with increasing distance to the footpoints (Lemmens and de Jager, 1986).

A well-studied case of evaporation is in the flare of 22 June, 1989 (Aschwanden *et al.*, 1993a), where the evaporating particles were apparently trapped in a long loop (68 000 km high). The individual electrons, moving at $v = 0.3c$ produced inverted U-type and type III radiobursts during their motion through the loop.

There were about 45 such injections, indicating that many energizations. They moved ahead of the evaporating plasma, that moved with 200 km s^{-1}.

The totality of the processes described here may be summarized under the name *Impulsive Phase Explosion.*

2.2.6. *Position of Sources*

After the discovery of the sequence of phenomena described here, a few crucial questions remain: that of the position of the source of energetic particles – hence that of the location of the initial process of energization; that of the nature of the energetic particles: electrons, protons or plasma, and that of the mechanism of their energization.

As to the location of the source of energization, Bogola *et al.* (1995) find for the flare of 12 November, 1980, that it was located in an area above the inversion line. For the flare of 30 April, 1980 (Figure 5) the area of primary energization seems to be located at some 10^4 km above the photosphere, near the apex of the flare loops. Further confirmation has been given by *Yohkoh* observations. These could first specify the average height of the source, this being 9700 ± 2000 km above the photosphere (Masushita *et al.*, 1992). Aschwanden *et al.* (1995c) confirm this height by deriving a lower limit of 7300 ± 800 km and Aschwanden and Benz (1995) found a height of 9200 ± 2300 km for the starting height of radiobursts.

In addition, it was found (Sakao, 1994) that the average time differences between the start of brightening of two corresponding footpoints is 0.0 ± 0.2 s. This can only be so if the source is situated between the two footpoints, not neccessarily exactly in the middle of a loop but presumably close to its apex. In view of the small time difference, in relation to the average length of the footpoints, the bombarding particles must be electrons: protons would move too slowly. Neither can the footpoints be heated by hot plasma. This is confirmed by Aschwanden and Benz (1995): in the thick-target situation the low-energy electrons should be delayed with respect to those of higher energies, and indeed: 50–100 keV electrons arrive 16.7 ± 1.9 ms earlier than the 25–50 keV electrons.

2.2.7. *Temperature and Energies*

The electrons should have an average velocity of $c/3$, the same value as that of electrons causing type III radiobursts. This confirms a study by Brown and Emslie (1987) who found that the observed spatial structure of the hard X-ray emission during the impulsive phase is best explained by electron heating. Piana *et al.* (1995), by inverting the thermal Bremsstrahlung spectrum found that during the impulsive phase of a flare there were two broad energy features, one peaking around 10^7 K, the other near 4.5×10^8 K. We tend to relate the high-temperature part with the hot kernels found by de Jager *et al.* (1987, cf., section 2.2.10 below). The temperature at the top of a loop has been measured a few times only (Takakura *et al.*, 1993; Wang *et al.*, 1994) and was found to range between 35 and 70 MK. A case of purely non-thermal emission in impulsive phase bursts was reported (flare of 27 June, 1980; Benka and Holman, 1992). The number density of electrons in the apexes of the loops was (5 and 2) $\times 10^9$ cm^{-3} (Takakura *et al.*, 1993).

2.2.8. *The Loop Model*

Takakura (1992) and Takakura *et al.* (1993, 1994) found that hard X-ray loops first brighten at their top, whereupon the X-ray brightening spreads to the footpoints. The observed velocities of about 10^4 km s^{-1}, however, do not agree with the above mentioned electron velocities and must relate to the motion of a plasma front. The succession of events was confirmed by Zarro *et al.* (1995) who also found that the hard X-ray emission is due to acceleration in a d.c. electric field, the strength of which increased from a preflaring value of 10^{-5} to 9×10^{-5} V cm^{-1} and remained constant during the impulsive phase. For a series of flares emitted from the same source region during the month of June 1991, Ramaty *et al.* (1994) could study the angular emission dependence. That could be interpreted in a loop model, with isotropically moving electrons in the top region, that emit gyrosynchrotron radiation, while precipitating electrons, emitting Bremsstrahlung, are moving down the legs. Aschwanden *et al.* (1993b) compared 100–3000 MHz radio observations and hard X-ray data, and found support for the view that there should exist a compact accelerating source, from where electrons are accelerated up- as well as downward, thus producing radio emission (the upward moving electrons) or X-ray emission (downward moving electrons). This result was confirmed in a later study of hard X-ray bursts by Aschwanden *et al.* (1995b), and this confirms an early suggestion by de Jager and Kundu (1963).

At this point we should remark that the above description is not a general rule: *'flares are different!'*. In particular we refer to the single-loop flares and the cusp-type flares (sections 6.6 and 6.7), but also in two-loops flares situations do occur where the loop-interaction takes place close to the footpoints; the Y-type interaction (section 6.3) being a clear example.

The next questions are about the (subtelescopic) structure of the interacting loops, and the process of energization, and: are we dealing with one, two or more loops.

2.2.9. *Magnetic Fields*

Some direct information on the *interaction process* can be obtained from an approach developed by Loran and Brown (1985) and somewhat modified by one of us (de Jager, 1986, p. 81). Basically, it is assumed that in a plasma the conversion of magnetic energy to other forms of energy cannot proceed on a time scale shorter than the Alfvén time, and that the escape rate of electrons accelerated at loop-interaction cannot exceed the limit set by the stability of the neutralizing current. Thus, one derives a lower limit for the magnetic field strength, values for the velocity of interaction and for the length-to-width ratio of the interacting region. For a number of flares studied by Loran and Brown one finds $B_{\rm min} = 100$ G, an interaction velocity of the same order as the Alfvén velocity, and an interaction region shape that has the same size in all directions, which esssentially means X- or Y-type interaction. Values for the magnetic field in a number of other cases (summary in de Jager, 1986) range between 100 and 1000 G, with preference for the lower values. Direct **B**-field measurements for the footpoints of the flare of 26 June, 1992 yields most probable values of 160 G and 400 G (Way *et al.*, 1994). Another typical value is the total number of energetic particles (> 10 keV), which ranges between 10^{34} and 10^{36} s^{-1} (Lin and Hudson, 1971; Hoyng *et al.*, 1976; de Jager, 1986b).

2.2.10. *Finestructure of Loops*

Flaring loops are inhomogeneous, with 'filling factors' of the order 0.01. This was discovered by Suemoto and Hiei (1959), and was noted later by many others. Švestka (1976, p. 75) summarizes the early evidences and gives more references.

Ichitomo *et al.* (1992) find for a limb flare of 2 December, 1991 a filling factor of 0.023. Hence, the fluxtubes must consist of many thinner tubes, the 'fluxthreads', a concept due to Tsuneta (1985) and de Jager (1985b), cf., also de Jager (1986a); de Jager *et al.* (1994). The fluxthread model finds support in a theoretical study by Koide *et al.* (1994), described in Section 3.4 of this review, where it is shown that upon footpoint rotation a fluxtube develops a current and return current system, situated in a coaxial thin shell, which may split in a number of helical thin fluxthreads (Figure 31). A natural consequence is that interaction between fluxtubes may lead to several fluxthread interactions, a phenomenon for which evidence is found in various flares. Observational confirmation that the process of *partial reconnection is a repeating and therefore ongoing process* in some flares is found in studies of the flares of 21 February, 1992 (Tsuneta, 1993; Tsuneta and Lemen, 1993) and of 2 August, 1994 (de Jager *et al.*, 1995).

Further support for these ideas may be found in the phenomenon of the *sub-second flares*, reported by Takano *et al.* (1994). They occur in groups and have a rapid rise and fall, with typical lifetimes of the order of 0.1 s. Such flares may yield essential new information on *elementary physical processes* in flares. These extremely brief bursts have some similarity to those studied by de Jager *et al.* (1987) in the flare of 21 May, 1984. During that flare a dozen of shortlived (0.1 s) peaks appeared within a total time interval less than 2 s. They are attributed to the formation and conductive decay of very hot, small kernels of flare plasma, that emit thermal gyrosynchrotron radiation. The temperature of these kernels was estimated at 300 to 500 MK, the size was about 350 km and the consequent conductive energy loss time was 60 ms, which is in excellent agreement with the observed decay time of the individual bursts. Model simulations showed that the ignition process can be attributed (Sakai and de Jager, 1991) to explosive coalescence of current-carrying flux tubes. In order to produce a dozen of such small kernels, there must have been a similar number of fluxthread interactions. Impressive support for the manifold interaction model is found in a study of $\simeq$ 6000 type III radiobursts (Aschwanden *et al.*, 1994a). The bursts appear in groups with on the average 13 ± 9 bursts per

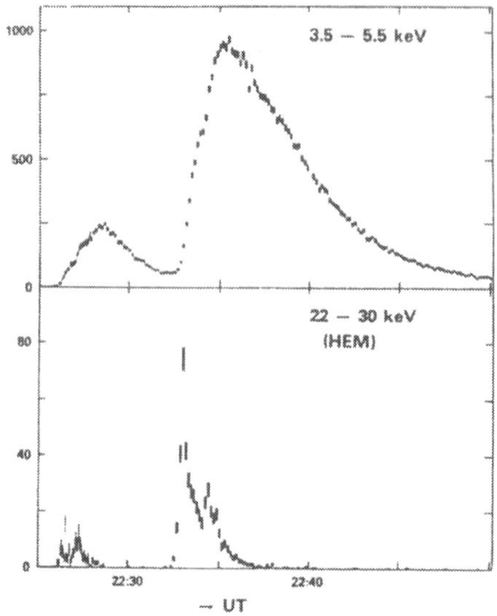

Figure 14. The impulsive and gradual phases, characterized by impulsive bursts in hard X-rays, and by the gradual variation in soft X-rays respectively, start at the same time. Flare of 5 November, 1980. (Duijveman *et al.*, 1982. From *Solar Phys.*)

group, a mean duration of 21 ± 14 s and a mean time interval between successive bursts of 2.0 ± 1.2 s. Comparison with Monte-Carlo simulations shows that the injection of electron beams is modulated either by a compact particle acceleration source, or by one that has a global organisation on a time scale of seconds.

2.2.11. *One or Two Loops?*

A further question: *one or two loops*? Assuming the filamentary structure of a loop, with its unstable helical current threads, the existence of one loop only is a sufficient condition for interaction between fluxthreads and consequent flaring. This can give rise to two footpoints, because theory predicts the occurrence of helical fluxthreads with opposite **E** fields. In a loop with only one **E**-direction, electrons can only move towards one of the footpoints, but in a configuration composed of fluxthreads with oppositely directed **E**-fields the electrons can stream down to both footpoints. Hence, the existence of two footpoints does not imply the existence of two loops, although the process of (partial) reconnection can evidently proceed more violently when two entire loops are involved.

Shimizu *et al.* (1994) drew attention to the class of *single loop flares*. They occur less frequently than multiloop brightenings. Takakura *et al.* (1993) described four cases. They started with loop-top brightening, progressing downward with $v = 10^4$ km s^{-1}. The initial loop-top temperature was 40–70 MK and the corresponding electron densities were $(5–2) \times 10^9$ cm^{-3}. Some of these are related to the cusp-type flares. Their characteristics, however, are such that we rank the single-loop flares among the confined/impulsive ones, in contrast to the cusp-type flares.

2.2.12. *The Gradual Phase*

It is important to realize that the *gradual phase* is *not* a sequential one, and that it does not merely succeed the impulsive phase when that phase has ended.

There are two evidences that may help in clarifying the gradual phase. First, as shows the example of Figure 14, the impulsive and gradual phases start in many cases *at the same time*; cf., the comparison of hard and soft X-ray intensity-time curves in Figure 14.

Next, the initial increase of the soft X-ray intensity curve demonstrates that for many confined flares their thermal energy content increases gradually till just after the last hard X-ray burst. This feature is also shown by Figure 14. Hence, energy is brought into the flare plasma by the series of hard X-ray bursts, and the gradual phase, which shows the time-development of the thermal energy content of the flare is due to the plasma being energized during the impulsive bursts and convected upward from the footpoints, and spreading from there over the flare area, merging with that from the other footpoint as it spreads, while cooling down by radiative losses. The thermal flare plasma can eventually occupy a fairly large volume and area and its cooling down process can take a long time, sometimes up to an hour, as appears from the duration of the gradual phase.

The observational physical *link* between the impulsive and gradual phases is that during the impulsive phase the thermal electron content of the flare increases monotonically, to reach a maximum value at the end of the impulsive phase – hence after the last impulsive phase burst. Thereafter it decreases (de Jager and Boelee, 1984). We note the consequent physical explanation of the 'Neupert effect' (1968) mentioned previously.

Noteworthy is also the gradual softening of the hard X-ray spectra over 50–200 keV at and after the maximum phase (Lee and Gary, 1994).

Conclusion: a high-temperature electron plasma is formed and injected into the flare area during the impulsive burst phase and at the same time the impulsive and the gradual phases start. The gradual intensity-time curve represents the time history of the flare's thermal energy content. The gradual phase ends only when the hot plasma has radiated its excess thermal energy. The above scenario was confirmed by Dennis and Zarro (1993).

2.2.13. *Summary*

We *summarize* the various evidences brought forward in this Section, while we neccessarily describe a 'standard' case, neglecting the exceptions. The confined flares occur inside closed magnetic coronal structures, that are only seldom disrupted by the flare. In those cases when disruption does occur, the confined flare may have triggered an eruptive one. Inside that closed magnetic structure the flare originates by some process of energization, very often by interaction of two loops but in some cases by instability in a single loop. The seat of energization is an area close to the top of the loop(s) involved, in many cases in a location, about 9000 km above the photosphere. Electrons are accelerated there, and bombard the lower chromosphere, which successively leads to the formation of the footpoints, to footpoint evaporation and to coronal explosions. The gradual phase starts at the same time as the impulsive one and its duration defines that of the flare. This phase is due to thermal gas emanating from the footpoints. It is significant that for a flare there is approximate equality of the energy contained in the > 15 keV electron beam, that of the convectively rising ‘evaporated’ (ablated) gas, and that of the thermal energy of the flare plasma in the gradual phase (de Jager, 1987). On the average, these energies are 10^{30} to 10^{31} erg. The average total number of bombarding > 15 keV electrons is 10^{38}.

2.3. Eruptive/Dynamic Flares; Observations and Physics

2.3.1. *Generalities*

The eruptive/dynamic flares normally occupy a fairly large volume in the corona near and inside spot-groups, in association with large coronal structures and/or filaments. Only very few of them occur outside active regions. They are impressive in sofar as their size and the associated phenomena are concerned: the phenomena in prominences like the *disparitions brusques* and prominence motion, elevation and ejection, and coronal phenomena such as expanding helmet structures and plasma ejection from the corona, such as Coronal Mass Ejections.

The highly improved (with regard to earlier spacecraft) soft X-ray imaging capabilities of *Yohkoh* gave rise to many interesting new studies on these complicated phenomena. *Yohkoh*’s soft X-ray telescope has unveiled the basic properties of the active regions, where we know that flares occur (Figure 1). These regions contain large numbers of loops and besides there are extended loops connecting active regions. The active region loops have average temperatures of $(5–8)\times10^6$ K and emission measures of $(5–7)\times10^{28}$ cm^{-3}. The average length, thickness, and electron number density are 33 000 km, 7000 km and $(7–8)\times10^9$ cm^{-3} (Shimizu *et al.*, 1992). In the aftermath of eruptive/dynamic flares the *post-flare loops* and the *giant coronal arches* are impressive features. The coronagraph aboard *Solar Maximum Mission* yielded many evidences of dynamic coronal features such as the Coronal Mass Ejections (CME’s), which are associated or at least correlated with eruptive/dynamic flares.

2.3.2. *An Eruptive/Dynamic Flare; Long Duration Events; Jets*

A fine example of the sequence of events in a *typical eruptive flare* is the limb flare of 21 February, 1992 (Tsuneta *et al.*, 1992a). A helmet streamer appeared in the corona, several hours before the flare, continuously expanding and restructuring. The expansion of the arch apparently created a suitable situation for the flare to start. The height increased and the footpoints separated from each other with velocities of 10 to 30 km s^{-1}. The morphology of the event shows that it was a combination of post-flare loops and a giant coronal arch. The flare lasted for many hours (L.D.E.: Long Duration Event).

L.D.E's are fairly common. In the latter part of the event they may show a cusp-shaped brightening at the loop-top, as described by Švestka *et al.* (1987), Tsuneta *et al.* (1992a), and Tsuneta (1994) (cf., section 2.3."). These cusps usually have a higher temperature than the other parts of the loops, and also in the decay phase the temperature in the outer arch was lower than in the middle part. It often appears that in dynamic non-impulsive soft X-ray flares the hottest part ($\simeq$ 20 MK) is located in compact regions on top of a loop (Acton *et al.*, 1992). These hot regions, with sizes as small as a few thousand km, form in the early rise part of flares, hence at flare onset or slowly thereafter, and they persist till well into the decay phase. The pre-existing material in the loop is being compressed, maybe due to the pinch effect of a strong electric current.

As is clear from the above, a combination of gradual and impulsive phenomena occurs often in eruptive/dynamic flares. A good example is the giant coronal arch of 21 May, 1980 (Švestka *et al.*, 1982a) and the one of 6 November, 1980 (Švestka *et al.* 1982b; Švestka, 1984). Sakai and Koide (1992) described a filament that started rising before the onset of an eruptive/dynamic flare. Nakajima (1994) investigated a gradual rise-and-fall flare with an impulsive burst superimposed. Triggering of the dynamic flare by a confined one seems to be the case here. Such observations are basic to the objection against classifying flares into impulsive and gradual ones. The designation eruptive/dynamic (or: open, in contrast to confined) is more suitable.

The *emission of X-ray jets* (Shibata *et al.*, 1992; Shibata, 1995; Yokoyama and Shibata, 1995) is a feature discovered by *Yohkoh*, that is sometimes, like in the case of Figure 15, associated with eruptive/dynamic flares. They are, however, not typical for them: confined flares may show them too. Typical jets have sizes between 5×10^3 and 4×10^5 km, average 1.7×10^5 km, with translational velocities of 30 to 300 km s^{-1}, average 190 km s^{-1}. Electron number densities range between 3×10^8 and 3×10^9 cm^{-3}. The analysis of the time development of the dynamic flare associated with the jet of 12 November, 1991 and of the related coronal magnetic field offers a way for diagnosing the magnetic field structure and its development (Figure 15).

2.3.3. *Cusp-Type Flares*

A *Yohkoh* observation that may clarify the explanation of eruptive flares is that of *cusp-type flares*. These consist of a loop with an enhanced apex part (Tsuneta *et al.*, 1992a; Tsuneta, 1994; Inda-Koide *et al.*, 1995). The observations show the interaction between a loop enclosed within an X-type field line system with the effects of reconnection at the X-line. An interesting case was described by Masuda *et al.* (1994): reconnection *above* a loop. That flare should rank among the eruptive/dynamic flares but the case is special since reconnection occurred first in the X-line area, which produced a downward moving plasma blob, whereupon a second reconnection at the loop apex was due to the collision of the neutral plasma

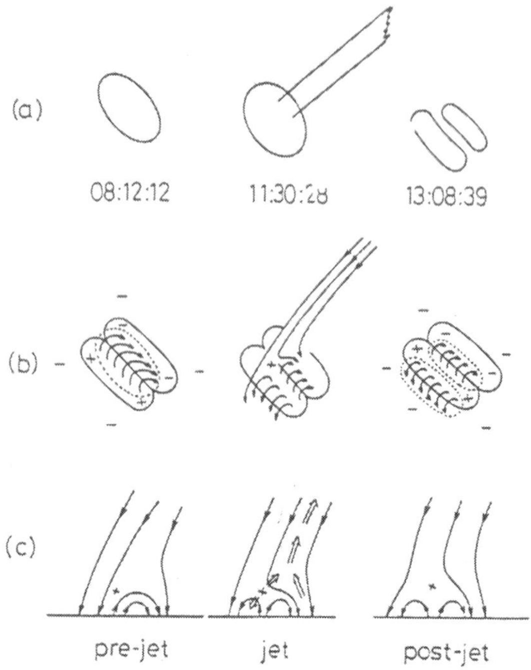

Figure 15. Schematic diagram of the possible development of a jet in the flare of 12 November, 1991. (*a*) Soft X-ray images before, during and after the jet. (*b*) Comparison with magnetographic data. (*c*) Side view of the magnetic reconnection process. (Shibata *et al.*, 1992. From *Publ. Astron. Soc. Japan.*)

blob with the loop, causing an impulsive flare with its cusp-type appearance. Possible explanations of this case are described in Section 6.7.3 of this review

2.3.4. *Post-Flare Loops and Giant Coronal Arches*

Eruptive/dynamic flares are often followed by rising systems of *post-flare loops*. There are two categories: the post-flare loops proper and the giant coronal arches.

The post-flare loops may reach heights up to about 10^5 km. The essentials of the loops after a flare of 6 November, 1980 are described by Švestka *et al.* (1987). We describe them here because they are typical. They were first visible in X-rays and later in Hα, after cooling and shrinking of individual post-flare loops (Forbes and Acton, 1996), while others developed on top of them. While individual loops shrank, the system as a whole rose over many tenthousands of km, over a period of some hours. Particle densities were up to 10^{12} cm^{-3}.

The observations by Seely and Feldman (1992), who found a case of upflow of a soft X-ray source with $v = 35$ to 90 km s^{-1} during the gradual phase of a flare, while the source moved from an altitude of 7000 to 80 000 km, refers certainly to the above-described first phase of the system of loops.

It may be that the post-flare loops are related to the long-lasting high-energy gamma-ray emission that is sometimes observed for a few hours during and after flares: we mention the 8-hours duration emission observed by Kanbach *et al.* (1993) and Akimov *et al.* (1993). The γ-rays are due to proton decay, while the pions are due to interaction between high-energy protons and α particles. Lau and Ramaty (1995) found that such long-lasting protons may be accelerated in turbulent twisted loops with sizes ranging from 20 000 to more than 100 000 km.

Dramatic examples of one of the many possible appearances of post-flare loops after an eruptive flare, are the *giant coronal arches* discovered by Švestka *et al.* (1982), by means of observations with the X-ray imaging facility (HXIS) in the *Solar Maximum Mission.* There are two categories: the *stationary giant arches*, that of 21–22 May, 1981 (Švestka *et al.*, 1982) being an example, and the *rising giant arches*.

The phenomenom consists of the appearance (or: the formation?) of a large soft X-ray emitting arch, with a lifetime of many hours to some days, and a typical length of a few times 10^5 km, extending up to and above 10^5 km above the photosphere. Repeated revivals appear to be possible (cf., Švestka, 1984). This discovery illustrates at the same time why the 'gradual' classification (in contrast to the 'impulsive' one) does not apply to this event because the first discovered giant arch originated after the two-ribbon flare of 21 May, 1980, with its well-developed impulsive phase. The phenomenon by itself had all the characteristics of an eruptive/dynamic flare. From the few observations of giant arches so far available it appears that the association with two-ribbon flares is not uncommon.

Yohkoh, with its 3 to 5 times better spatial resolution than SMM gave new stimulus to the study of the rising giant arches (Švestka *et al.*, 1995). As a rule they succeed the eruptive/dynamic flares and rise with speeds of 1.1 to 2.4 km s^{-1}. The upward velocity stays constant within 5 to 24 hours. Temperatures in the brightest parts of the arches, as measured for an arch on 2–3 November, 1991, are of the order of 2 to 4 MK. Particle densities range from 10^{10} cm^{-3} at 150 000 km to 10^9 at 250,000 km height, an altitude reached 11.5 hr later. The event of 21–22 February, 1992 is interesting because it shows both a loop system, rising to 120 000 km, and an arch, emerging from behind the loop and rising to 240 000 km.

2.3.5. *Arcades of Coronal Loops*

Related to the theme discussed in this section is the discovery of large *arcades of coronal loops* visible in X-rays (cf., Tsuneta *et al.*, 1992a), with the *walking arcades* as a special sub-category (Nakajima, 1994; Kosugi, 1994). Tsuneta (1992b) describes the disappearance of a polar crown filament, followed by the appearance of a closed loop arcade. Motions in the arcades have velocities of the order of 30 km s^{-1} while the associated prominences have velocities up to and above 100 km s^{-1} (Nishio *et al.*, 1994). Interaction between the loops of the arcade is not uncommon. Watanabe *et al.* (1992) observed a quiescent filament on 28 November, 1991 and a rising one on 7 November, 1991. Both were associated with a soft X-ray arcade that appeared 2 to 3 hours after onset of a flare. Nakajima (1994) descibed a long duration flare (> one hour) associated with an arcade and a large loop seen near the solar limb (Hanaoka *et al.*, 1994).

2.3.6. *Global Structural Changes; the Grand Instability*

Motions of loops can be followed by restructuring of the loops, apparently through mutual interaction. Restructuring takes place on length scales ranging from the size of an active region to the solar radius. The associated filaments may undergo dramatic changes, diappearances or eruptions. Hence, filament eruption can be related to, or can trigger, arcade formation (Kosugi, 1994).

There are hence connections between eruptive flares and coronal motions or restructuring. Tsuneta *et al.* (1992b) describe the dissappearance of a coronal crown filament, followed by the appearance of a closed loop arcade that moved westward (a 'walking arcade'). Thereupon a cusp structure appeared at the solar limb. This *global structural change* was ascribed to non-explosive ongoing reconnection. It makes sense to distinguish between 'quiet' and 'active' active regions (Uchida *et al.*, 1992). While the corona may expand occassionally above active regions, this happens always above 'active' active regions. During the expansion, with velocities between a few and some tens of km s^{-1} the overall structure of the corona remains preserved. The phenomenon must be due to some form of energy injection into the coronal part of the active region.

The review of the various observations related to eruptive/dynamic flares and their many associated phenomena in and around active regions suggests that a flare is neither the primary event, nor the cause of these many events. All these events should instead be considered as being elements of a *grand instability* associated with some degree of *restructuring of the active region*.

2.3.7. *Model and Outline of Theory*

It is not surprizing that, in view of the diversity of the observations, different theories have been developed for explaining the eruptive/dynamic flares. But in spite of this diversity, every theory should in any case be able to explain the common feature of all eruptive flares: filament activation and gradual radiative energy release, associated with dynamic coronal activity and Coronal Mass Ejections. The observations and inferences presented here do not lead to one typical configuration that would do justice to all observations. Instead, the diversity of dynamic flares requires a similar diversity of 'typical models'. Yet a simple general picture may be put forward. It is based on considerations and models advanced by Syrovatsky (1966) and others for describing the opening of the field-line system, followed by reconnection, as has been worked out for the flare of 6 November, 1980 by Švestka *et al.* (1987). It consists (Figure 16) of a magnetic configuration with an X-line at some height above the photosphere and an 'open' configuration higher-up. Inside this configuration one single or two interacting loops may be situated. A pinch-sheet may develop at the X-line. As a consequence of the reconnection

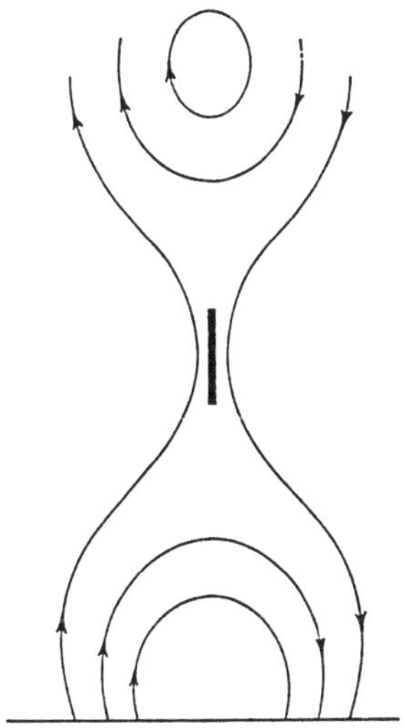

Figure 16. Schematic of the magnetic configuration in an eruptive/dynamic flare.

several things may happen at nearly the same time: The – initially open – field lines shrink after reconnection (by about 20 to 30%, according to observations by Forbes and Acton, 1996). The X-line rises upward. A cusp forms above the loop, between the X-line and the loop-top. The pinch-sheet instability is shown by the rise of the filament and by helmet structure variations, and often starts before the loop instability results in a flare. The consequent changes of the field line system above the pinch-sheet can result in a Coronal Mass Ejection.

The next problem is that of the start of the instability; hence: *why is all this happening*?

One of the first theories, applicable to dynamic/eruptive flares, that was successful in this respect, was developed by Kuperus and van Tend (1981) based on previous work by van Tend and Kuperus (1978). Basically, the process is assumed to be initiated by the increase of a current embedded in the background magnetic field of an active region. The current increases because of footpoint motions, an effect that might (but should not neccessarily) be related to the emergence of new flux. It is also possible that only one of these two mechanisms contributes to the increase of the electro-magnetic energy content of the configuration. This energy, being built up by the mechanical action, is stored in the loop system as long as such is possible. The consequent increase of the current-system causes an expansion of the magnetic field which manifests itself as a slow increase of the size of the current-system, but it may eventually result in more dramatic consequences: the increase of the non-force free current-system can make the filament rise and explode, depending on the current-intensity. This may result in the transformation of the field-system into a configuration that is open towards interplanetary space. The process develops on the Alfvén time scale of the filament, and the theory seems capable of explaining the phenomena observed in eruptive/dynamic flares and the start of the associated plasma cloud emission.

The theory is supported by the observation that filament rising often begins before the flare, as observed in X- and γ-ray emissions. The motion of the filament continues during the flaring period. This shows that the gradual change of the current system (as manifested by the filament) is the primary phenomenon. In that line of thought, also following the route set by Kuperus and Van Tend (1981), Mikic *et al.* (1988) studied the evolution of a solar coronal magnetic field subjected to shearing photospheric motions. For motions exceeding a certain threshold the bipolar magnetic field becomes unstable, which results in the formation of a current-sheet above the polarity inversion line. This instability causes upward plasma flow.

In an alternative model Uchida and Shibata (1988) developed a theory for lifting chromospheric material into the loop tops, where it is heated and confined, leading to upward motion of the loop. The model, however, disagrees with the observations, that show that the individual loops do not expand; they rather shrink (Švestka *et al.*, 1987; Forbes and Acton, 1996), while the system of loops as a whole expands, through the formation of new loops.

After having described *why* an eruptive/dynamic flare developes, the next problem is *how it happens and develops*.This question has been elaborated by many authors, leading to the neutral sheet-type model as presented in Figure 16. Apart from a number of contributions to conference proceedings which were useful in developing ideas, the main stepping stones were laid by Syrovatsky (1966), Hirayama (1974), Kopp and Pneuman (1976), and Sturrock (1989). In these ideas the field first opens, which causes the filament to rise, whereupon reconnection starts, beginning with the development of the eruptive/dynamic flare. Further research, by Sakai and Koide (1992a), showed that a current-sheet extending along the X-point of the magnetic configuration becomes thinner (horizontally) and more elongated during the growth of the current-system. The sheet, identified with the filament, rises with a calculated velocity of about 100 km s^{-1}, which is in good agreement with the average observational value. Another interesting fact is that the theory predicts filament rising *before* the start of the flare, which indeed often happens in eruptive/dynamic flares. The characteristic time for the process is the Alfvén time, estimated at about 300 s under typical conditions. The theory was claimed to be developed for confined/impulsive flares but it applies better to the eruptive/dynamic ones. Takakura (1992) presented a scenario for single-loop instability based on heat input near the loop top. The consequent development of plasma waves increases the electric resistivity, which is followed by loop and footpoint heating.

A possible scenario for cusp-type flares, by Sakai *et al.* (1995), is based on the fact that approaching field systems in an active region may lead to plasmoid ejection (the 'melon seed effect') as shown in Figure 17. The downward moving plasmoid interacts with a lower situated loop and thus the cusp-type feature originates (cf., section 6.7).

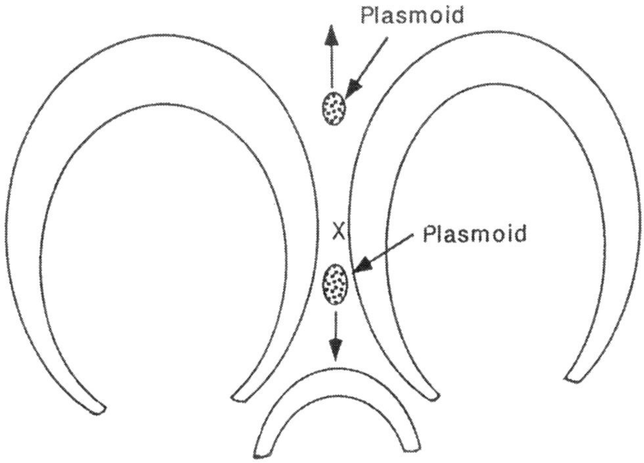

Figure 17. Plasmoid ejection in approaching field line systems; the *melon seed effect*. Ejection from the X-point. (Sakai *et al.*, 1995. From *Solar Phys.*)

2.4. Emission of Plasma and Energetic Particles from the Sun

In the previous sections mention is made of various kinds of plasma and energetic particle emission from the Sun: the Coronal Mass Ejections (CME), the emission of Solar Energetic Particles (SEP), which includes both the gradual SEP's and the impulsive 'proton events', and the 'gama-ray/proton (GR/P) flares'. We remarked that:

– the CME's are correlated with eruptive/dynamic flares, not with confined ones;

– for those SEP's that are related to confined flares the particle fluxes are generally low; they are higher for SEP's related to eruptive flares;

– GR/P flares are often related to eruptive/dynamic flares;

– the confined flares are related to abundance anomalies in interplanetary particle fluxes, most prominently the He^3 anomaly.

2.4.1. *CME's*

The CME's were discovered in *Skylab* observations. They are seen in Thomson-scattered white light observations (Figure 18), and are generally observed by K-coronameters in space, sometimes also on Earth. A few thousands of them have been observed sofar (Burkepile and StCyr, 1993; Dryer, 1994). The average mass is 5×10^{15} g, and the velocity ranges between 100 and a few 1000 km s^{-1}. Often, they are discontinuously accelerated after their start. In space they are often preceeded by an interplanetary shock. The shock-strengths depends on the interplanetary field configuration (Gosling, 1993).

The CME's are often associated with disrupting, rising or eruptive filaments, with coronal helmet streamers (Harrison, 1991), with arcades of fluxtubes (Uchida,

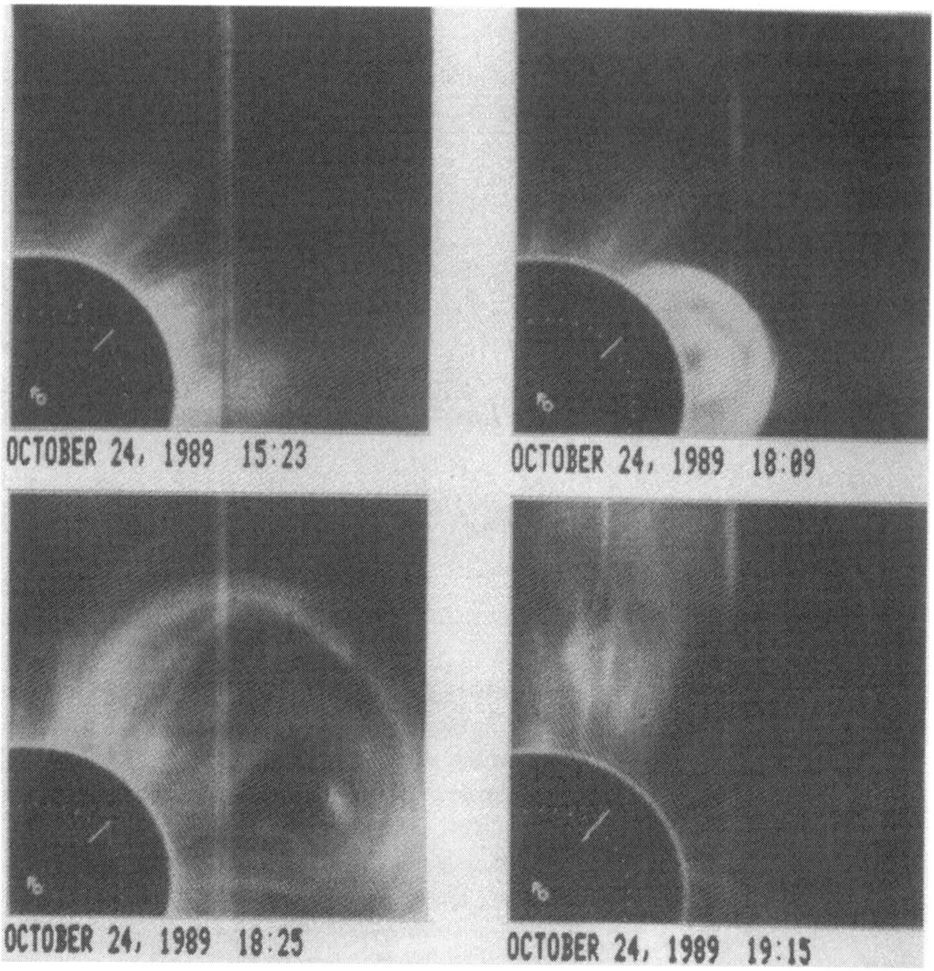

Figure 18. Coronal Mass Ejection related to the Ground Level Effect of 24 October, 1989. The CME speed was 1450 km s^{-1}. (Kahler, 1994. From *Astrophys. J.*)

1995). The relation with sudden disappearances of filaments exists only for *dynamic sudden disappearances* (Mouradian et al., 1995). These consists of an expansion followed by ejection of prominence plasma. This, in turn, is followed by a CME. St. Cyr and Webb (1991, reviewed by Dryer, 1994) found the following correlations:

- with eruptive prominences: 0.88;
- with Hα flares: 0.34;
- with X-ray events: 0.63;
- with radiobursts of types II or IV: 0.23.

The CME's are also associated with, and often arise from, pre-existing coronal streamers (which implies that the field pattern already existed before). The streamers are observed to 'swell up' during one or a few days before the CME event (Webb *et al.*, 1994; Feynman and Martin, 1994). Broadly speaking, the CME's are related to rearrangement of plasma with respect to the inversion line. All these associations refer also to the eruptive/dynamic flares. The CME's are indeed related to such flares, but not very strictly: about half of them occur before, the other half after a flare, with time differences that may amount to one or several hours (Harrison, 1991; Harrison and Sime, 1992). There is a danger, though, that such statistics is spoiled by an incorrect association of CME's and flares.

Although CME's are definitely related to eruptive flares, it sometimes seems that they are related to confined flares, but that only seems so, as a consequence of a combination of two features: (a) certain eruptive/dynamic flares include events outside strong magnetic fields, that do not cause a chromospheric flare, and (b) eruptive flares can be triggered by confined ones (Švestka, 1996, private communication).

Actually the three features, CME's, long enduring flares and dynamic filaments are all part of a complex of phenomena, that may be unified under the heading *eruptive/dynamic flare*. Energy-wise, the CME is the more important feature; the flare is secondary. The energy contained in Giant Coronal Arches is comparable to that of the CME's (Švestka, 1996, private communication). The CME's and the Giant Arches seem to be counterparts of respectively the rising and stationary type IV radiobursts. The impulsive high-energy particles ('proton events') observed during CME's are however produced by the flare, and this observation shows the importance of flares as seats of fierce acceleration. Pérez Enriquez and Mendoza (1995) summarize: *The role of the flare is to produce and accelerate the particles; that of the CME to (...) open the field lines and that of the CME shock to further accelerate and transport the particles in longitude towards the central meridian.* According to Kahler *et al.* (1994) the CME does lead to the acceleration of gradual relativistic particles.

Watanabe *et al.* (1992) describe the filament of 28 November 1991, and the eruptive prominence of 7 November, 1991, that were both associated with an *arcade* seen in soft X-rays. The arcade appeared 2 to 3 hours after the onset of an Hα event. At the time of the flare the prominence had already reached a distance of $0.3R$ from the limb. The phenomenon suggests reconnection between structures visible in Hα. Another case is the arcade of 4–5 November, 1992, that developed later into a large CME. The Giant Arches are part of this process too: as the CME disrupts the corona, the active region loops, left behind, may reconnect (Šimberová *et al.*, 1993), and they begin to expand into the coronal area left behind by the CME (Švestka *et al.*, 1995). The flare of 12 March, 1989 had an impulsive main part and a 2-ribbon peripheral part. Mass ejection occurred at flare maximum, from a rising filament that moved with a line-of-sight velocity > 600 km s^{-1} (Den and Kounenkov, 1993).

Summarizing, there exists a group of correlated events: the flare, the filament instabilities, and the CME. They are due to a *grand instability* in an active region. Gosling's (1993) statement about the 'flare myth', decoupling flares from CME's, is in its general form not correct; see also Švestka (1995), and Gosling and Hundhausen (1995).

2.4.2. *Solar Energetic Particles and Proton Events*

SEP fluxes are detected on Earth by ionization chambers (since 1933), neutron monitors (since 1953) and in space by particle detectors since 1965 (see the review by Stoker, 1995). Proton events are known since 1942, but only the relativistic events. After 1957, space instrumentation allowed for the observation of Solar Proton Events of lower energy; they occur at an average rate of 70 per solar cycle (Smart and Shea, 1989). Of the events observable at the ground level, called *relativistic solar proton events*, Stoker (1995) lists 40 events observed between 1966 and 1992. The correlation between gamma-ray flares and proton events, leading to the notion *GR/P flare*, is only known since 1980, when systematic γ-ray monitoring started. In Antarctic snow layers the occurrences of solar protons in the Earth's upper atmosphere has been traced down to the large flare of 1909. The effect is measured by the enhanced concentration of NO_3 ions, due to stratospheric ionisation, in snow layers (Dreschhoff and Zeller, 1990; Dreschhoff *et al.*, 1993). Solar proton events with large fluences are associated with solar activity near the central meridian of the Sun; those with large proton fluxes with activity near the Sun's western limb (Shea et al., 1993). For those SEP-events that are associated with Hα flares, the flares are situated at the feet of helmet streamers, and they are related to CME's (Pérez Enriquez and Mendoza, 1995).

It makes sense to distinguish between the (gradual) SEP's and the (impulsive) Proton Events. With regard to the emission of the gradual SEP's and the Proton Events the situation is as follows (Cane *et al.*, 1988; Kahler *et al.*, 1994; Kahler, 1994; Reames, 1993; Pérez Enriquez and Mendoza, 1995). The energetic particles in most major Solar Proton Events do not come from the flare itself, but from shocked plasma driven by a coronal mass ejection. The source region of the gradual SEP's may be situated at some 10 solar radii from the Sun (Kahler, 1994). The energetic particles in relativistic GLE's (Ground Level Effects) are accelerated closer to the Sun, at about 2.5 to 4 $R_{\odot}$ (Cliver *et al.*, 1982; Kahler, 1994). This makes clear that such particles are byproducts of dynamic flares, accelerated in flare-driven shocks.

In some contrast to the gradual SEP events stand the more impulsive 'Proton Events'. These energetic particles are related to confined/impulsive flares; they are electron-rich, have large enhanced He^3 as well as smaller enrichments of heavier nuclei, like Fe. Such particles appear with smaller fluences than those related to eruptive/dynamic flares. We note that this is understandable, because the field configuration of a confined flare hampers the escape of flare particles.

With regard to the GR/P flare, a preambulary remark is that there is no *essential* difference between gamma-ray emitting flares and others. The ratio between 4–8 MeV and $>$ 300 keV fluences are the same (Cliver *et al.*, 1994). Gamma-ray flares are simply the more powerful ones. Contrary to this property, the experience of the last decade has shown that there is no strict correlation between gamma-ray flares and proton events. The flare of 27 April, 1981 had strong gamma-ray emission but no related protons. Pesses *et al.* (1981) mention non-gamma flares for which the number of interplanetary protons was much higher than those for the average gamma-ray flare.

Bai *et al.* (1985) found that two classes of GR/P flares exist: the *gradual* ones have a gradual variation of the hard X-ray fluxes, a long lasting soft X-emission, extended coronal loops, coronal mass ejections. They emit intense fluxes of energetic interplanetary protons. The *impulsive GR/P flares* have the opposite behaviour. In both classes of GR/P flares the protons are emitted during the first phase of the flare, apparently during the impulsive phase. In the gradual GR/P flares further acceleration of energetic particles occurs in a later phase, in an 'open' field line configuration from where particles can escape into space. We note the similarity to the gradual SEP events.

The two-ribbon flare is a special case of the eruptive/dynamic flares; it is an instability in an *arcade of flux tubes* and involves strong magnetic fields. Already for that reason it is a very energetic phenomenon, sufficiently energetic to open an extended field line system. The field line system of such a flare has the shape of an inverted Y; from it particles may escape to space. On the average, the magnetic energy available in a two-ribbon flare is about five times larger than in the average flare.

It is remarkable and undoubtedly very significant, that as a rule most of the energy of the flaring region is imparted to the ejected particles and plasma. Canfield *et al.* (1980) summarized the energetics of the flare of 5 September, 1973. In units of 10^{30} erg:

- total radiated energy: 4
- erupted prominence: 1
- emission front: 3
- shock wave: 6
- magnetic energy: 10.

In the same units the total *net* energy of a typical active region that produces an eruptive flare, (i.e., the difference between the force-free and potential energy) is of the order 1000 to 10^4 (Wu *et al.*, 1993a, b), which is enough to power a large number of flares in the active region. This also makes clear that even the appearance of a number of flares in one active region will hardly change the overall magnetic topology of the region. But: *'flares are different'*: Sakurai *et al.* (1992a) compared magnetograms before and after the flare of 6 February 1992 and found that a current of 2×10^{11} A, existing before the flare, had disappeared after the flare. This corresponds with a total energy loss of 7×10^{31} erg. Dryer (1994) found that the total energy of a CME cannot be stored in the pre-existing corona: he suggests that it comes from the smaller-scale reconnection processes that cause the flare.

The above summary shows that the emission of energetic particles (SEP) and of plasma (CME) is a phenomenon related to eruptive/dynamic flares, practically not to confined/impulsive ones. This has to do with the field line configuration. Any flare would have the possibility to emit accelarated particles into space when the field line system would be open towards space. From confined flares the particles can only excape by a diffusion process.

2.4.3. *Summary*
It seems now clear how the three classes of escaping plasma or particles mentioned in the early part of this section should figure in the classification confined vs eruptive.

Confined flares are impulsive, and occur in a closed magnetic topology. Energetic particles are generated during their impulsive phase but only a small fraction of them can leak to interplanetary space. Actually, the 'leaking away' was essential for the discovery of the He^3/He^4 anomaly. If the confined flares would have been rigorously 'closed' it would have been difficult for particles generated there to escape. The confined/impulsive flares are responsible for the (impulsive) Proton Events and are related to the impulsive CR/P flares.

The *eruptive/dynamic flares* are correlated with the emission of plasma and energetic particles from the Sun, but they are not neccessarily their cause. In about half the number of cases CME's preceed a flare; in the other half they follow the flare (Harrison, 1991; Harrison and Sime, 1992). It seems therefore that *a grand instability in a large X-type magnetic topology* causes flare, filament instability and CME. The related gradual Solar Energetic Particles are first accelerated in the flare and thereafter in the shock-wave field associated with the early phases of the CME-ejection. Any theory of eruptive/dynamic flares should take these *four aspects* into consideration.

In Section 7 we will return to this matter in our discussion of the phases of acceleration in and after flares.

2.5. Observational Demands to Theory

In the review of observational data presented in this section we conclude that the main difference between the confined/impulsive and the eruptive/dynamic flares lies in the electromagnetic configuration, which may be characterized by the concepts: 'closed' or 'open'. The demands observations put to theory are the following. One should study the physics of single loops, how they are structured and by what mechanism they become unstable. One should examine if such loops can give rise to flares. Next, one should study the interaction between two fluxtubes in order to verify the hypothesis of two-loops interaction. The possibility of single-loop instabilities is a challenging one and needs further study. The hypothesis that confined/impulsive flares occur in a magnetically closed and eruptive/dynamic ones in a more open configuration seems now well established. The formation of current sheets/filaments in active regions and their lifting needs study. Finally, an important topic is the 'grand instability' of the erupive/dynamic flare complex including the active region, in relation to the various phenomena that occur in connection with eruptive/dynamic flares. One should hence endeavour to explain in one model the emission of solar plasma and/or energetic particles, the occurrence of flaring instabilities and the associated phenomena such as the rise and disappearance of filaments.

In all configurations and phases mentioned here, acceleration of particles and plasma can and will occur. Their explanation is another challenge to theorists.

Reversely, observational means exist to check several of the theories related to flares, particularly the assumption that the two kinds of flares are related to different magnetic topologies. The quality of present-days instrumentation is such that the reality of this hypothesis can be verified observationally. There exist sophisticated radio-astronomical telescopes and high-resolution optical instruments. *Yohkoh* is a powerful spacecraft that is well suited to the study described here.

In the next three Sections we develop the theoretical aspects of single loops and their instabilities as well as of multiple-loops interaction and instability. In Section 6 the results of the theoretical considerations will be compared with modern observational data, and we will investigate if the various theoretically developed ideas on flare typology find their counterparts in reality. In Section 7 we discuss aspects of the three phases of acceleration in flares, and we compare them with observations. In Section 8 we summarize the results of this review.

3. Current-Loop Generation and Coalescence Dynamics

3.1. PHOTOSPHERIC MOTIONS: ROTATION, SHEAR MOTION, AND COLLISION

It is nowadays well recognized, from various high-resolution observations and theoretical studies, that magnetic fields play an important role in various solar phenomena such as flares, prominences, spicules and coronal heating. Among these phenomena, the flares are the most energetic (and very complicated) transients. For previous summaries of solar flares, see Švestka (1976), Sturrock (1980), Priest (1985), Sturrock *et al.* (1986), Tandberg-Hanssen and Emslie (1988), Zirin (1989), Parker (1994).

It is generally accepted that a flare is a manifestation of the explosive release of magnetic energy stored in the corona, associated with various ways of nonthermal energy dissipation, such as explosive plasma heating, acceleration of charged particles up to relativistic energies, and yielding the production of electromagnetic radiation in the whole electro-magnetic range, from radio wavelengths to γ-rays.

The magnetic energy, released during a flare, is assumed to be generated in the convection zone, and can be built up gradually in the corona as a current-system due to the photospheric motions. Magnetic stress can also be stored in the corona, when a magnetic field emerges from the photosphere and interacts with pre-existing fields (Heyvaerts *et al.*, 1977; Forbes and Priest, 1984; Shibata *et al.*, 1992; Yokoyama and Shibata, 1994). The photospheric motions are generally very complicated, but as shown in Figure 19, these motions can be idealized as rotation of the feet of

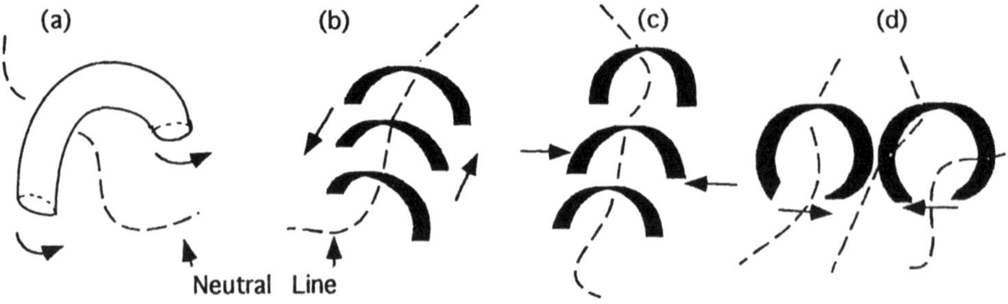

Figure 19. Schematic pictures showing current generation processes: (a) twisting motions near the footpoints; (b) shearing motions; (c) compression of magnetic arcades; (d) collision of two loops. Dashed lines are magnetic inversion ('neutral') lines.

magnetic flux tubes, as shearing motion, as compression of a magnetic arcade, and as collisions of loops and arcades.

The current-sheet formation and catastrophe theory in a sheared arcade were discussed with applications to two-ribbon flares by Zwingmann (1987) and Forbes and Isenberg (1991). Current-sheet formation was discussed for the case of colliding arcades by Low and Wolfson (1988) and Vekstein and Priest (1992). The dynamical evolution, causing rapid plasmoid formation in a pair of sheared arcades (Mikic *et al.*, 1988) as well as in a triple of sheared arcades (Biskamp and Welter, 1989), has been investigated, using MHD simulation codes. Parker (1994) presented a first extensive treatment of an MHD theory of spontaneous magnetic discontinuities (current-sheets). He suggested that *nanoflares* occur to some degree in all cosmic magnetic fields that are extensively deformed by convection. He also proposed to consider a flare as a firestorm of nanoflares that continues until the magnetic energy is depleted to such a level that local tangential discontinuities no longer exceed the threshold for the onset of explosive reconnection.

Here we concentrate on the current-generation process due to twisting of a coronal loop by rotational motions of the feet of the loop. It is shown in the next section that the radial profile of the rotational velocity at the feet of a loop is crucial for determining the radial profile of the current-density induced by the rotational motions. Futhermore, the stability of a current-loop can be determined from current-profiles as well as from the photospheric line-tying effect (Hood and Priest, 1981; Mikic *et al.*, 1990). As pointed out by Biskamp (1993, p. 340), the current-profiles in a coronal loop are different from those in a Tokamak plasma and the line-tying effect may lead to differences in the physical processes of the magnetic energy release between a Tokamak plasma and coronal loops. Following Beaufume *et al.* (1992), we also emphasize here the importance of the stability (Koide *et al.*, 1994) and the nonlinear evolution (Koide and Sakai, 1994) of a loop with a shell-current structure as well as a closure current, which can be naturally induced by the photospheric rotational motions (Sakai and Fushiki, 1995).

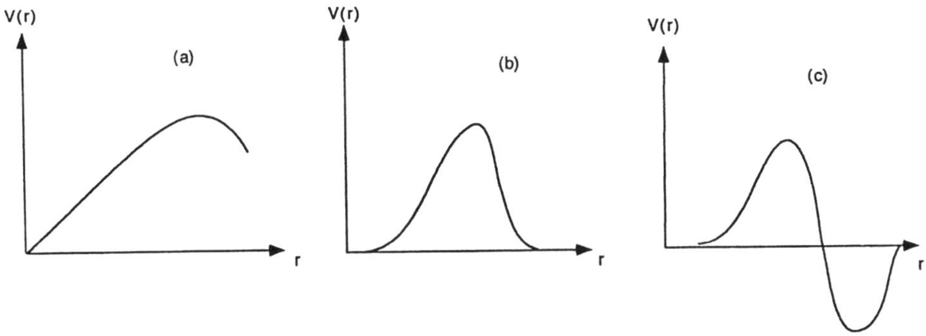

Figure 20. Schematic profiles of rotational velocities at the footpoint of a loop: (a) quasi-rigid rotation; (b) rotation with a shell-structure; (c) rotation with shearing motion.

3.2. Loop Generation with a Shell-Current

Spicer (1982) and Melrose (1991) gave a qualitative discussion of current-patterns in the solar corona, associated with solar flares. Melrose (1991) classified the coronal current-patterns in two types; neutralized and un-neutralized coronal current-patterns. For a neutralized coronal pattern there is a direct and a return path, both through the corona, so that the net current crossing the surface above the photosphere remains equal to zero. In contrast to this is the case when the coronal current flows from one footpoint to another with no return current above the photosphere. This is referred to as the un-neutralized coronal current-pattern. Present-days observations of the vector magnetic fields in the photosphere suggest that there is a net current flowing through the corona along coronal magnetic loops from one footpoint to the other (case of an un-neutralized coronal current-pattern). Melrose (1991), however, gave two reasons for the possibility that an existing return-current is not detected: One reason is that the return-current channel may be broad and has weak currents, that escape detection. The other is that the return-current flows in many thin current-channels (the 'fluxthreads') throughout the current-carrying region, which cannot be detected by present-days observational means.

In this section we discuss how neutralized current-patterns can be generated from a twisting motion of one footpoint relative to the other. In a cylindrical (uncurved) loop model (with radius R) for a field that is being twisted, the current-distribution across the cylindrical loop depends on the radial profile of the azimuthal velocity v_ϕ of the plasma flow at the photospheric level. Here, we consider three types for the azimuthal velocity pattern, as drawn in Figure 20. The rotational motions are applied at one or the two ends of the loops, simulating footpoint twisting. The first case is that of nearly rigid rotation near the axis of the loop, which induces a current with maximum strength at the center of the loop. In the second case the azimuthal flow is confined in a shell, in such a way that v_ϕ has a maximum at some distance from the axis of the cylinder. In that case the sign of the current changes at that radial distance R, where the azimuthal velocity has its maximum, which means that there appears a return-current outside the loop. In the third case the azimuthal flow changes sign at a distance R from the axis. Then there appears a current-loop with three coaxial shells of which two are direct-current shells with a return-current shell in between them. Here we note that the above three types of the neutralized current-loop patterns can also be realized by a shearing motion, which may be due to a displacement, parallel to the neutral line, of one footpoint relative to the other. The second case may correspond with a current-structure having a neutralized arcade pattern, while the third case may correspond with an arcade structure containing three current-layers, consisting of two direct currents and one return current in between them.

We discuss in some detail the second case with both a direct and a return shell-current, a case obtained from three-dimensional ideal MHD simulations (Sakai and Fushiki, 1995). A 3-dimensional ideal MHD code was used (Fushiki and Sakai, 1995), in which the numerical scheme is based on the two-step Lax–Wendroff method. The system sizes, L_x, L_y, and L_z in the x, y, and z directions, respectively, are $0 < L_x = L_y < 2\pi L_0$, and $0 < L_z < 4\pi L_0$, where L_0 is the characteristic length scale. The number of mesh points in the x, y, and z directions are 80, 80, and 50, respectively. The time step is $0.0015\tau_A$, where τ_A is the Alfvén time. The boundary conditions are free for all directions. The plasma beta is taken unity.

As the initial condition, we impose a rotational velocity at both footpoints of the loop, which has no initial current. The plasma pressure is chosen such that balance is maintained with the centrifugal force produced by the rotational motions at the footpoints. To simulate the photosphere, where the rotational motions occur, we increase by a factor ten the density of the region where the twisting motions are imposed, in relation to that of the region without rotational motions. Of course this density enhancement is far from the reality of the actual solar atmosphere, but the mass loading effect near the area of rotational motions is quite important for maintaining the current-generation and for enhancing the intensity of the current-density.

The magnetic field is initially taken homogeneous in the z-direction, and the amplitude of the rotational velocity is taken $0.3c_s$ (c_s is the sound speed) in its maximum. A functional dependence of the velocity on r and z was assumed:

$$V_\theta(r, z) = V_\theta(z) r / [(r-1)^2 + (0.5)^2],$$

where

$$V_\theta(z) = 0.5 V_m [\tanh(z-\pi) + \tanh(z - 3\pi)].$$

Figure 21 shows the time development of the current J_z along the magnetic field, which is generated due to the rotational motions at the loop footpoints. As seen in the figure, a direct current with a shell-structure can be produced, starting in the region of the twisting motions and propagating towards and into the center

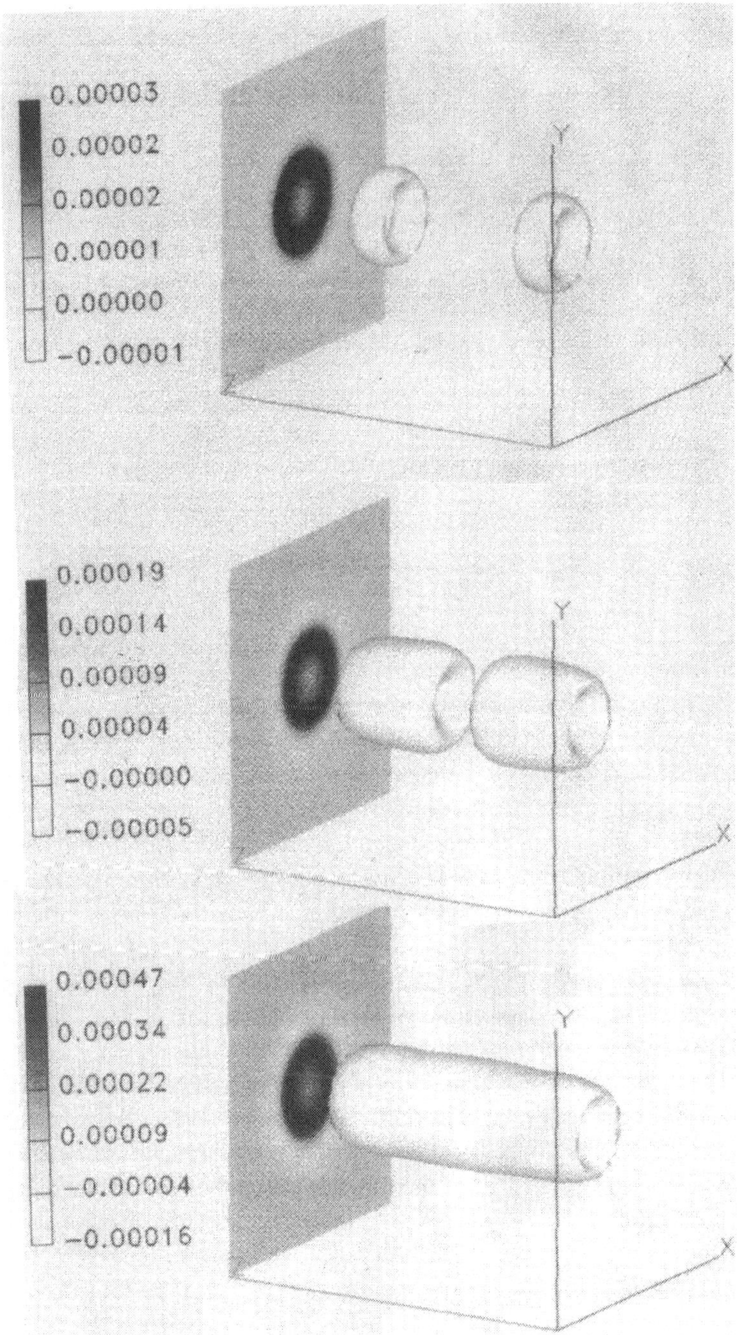

Figure 21. Time history of generation of direct current J_z. From *above* to *below*: t = 0.3, 0.9, and $1.8\tau_A$. The gray-scale array shows the current J_z in the plane $z = 4\pi L_0$. Two current-systems originate, that are connected later. (Sakai and Fushiki, 1995. From *Solar Phys.*)

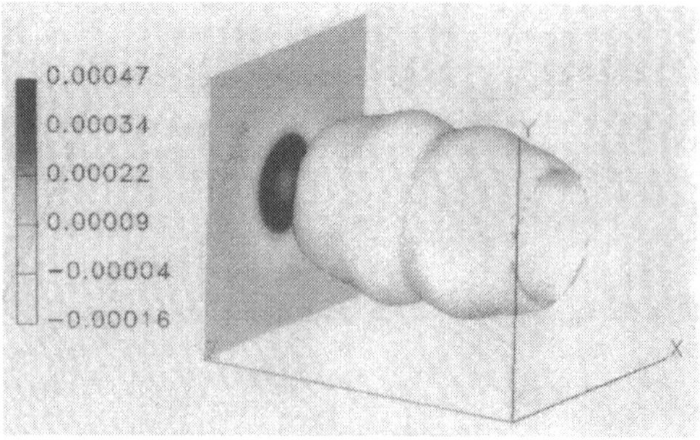

Figure 22. Return current distribution at $t = 1.8\tau_A$. This diagram is to be compared with the lower panel of Figure 21. (Sakai and Fushiki, 1995. From *Solar Phys.*)

of the loop. As a result of the mass loading in the area of the twisting motion, the current-generation remains continuous.

For checking the importance of the mass loading, the calculation was repeated, but without introducing mass loading in the area of the twisting motion. In that case the current-generation appears to be weak and it does not continue for a long time. Therefore the mass loading effect is important for maintaining current-generation and also for generating a high current-density. In the present simulation the density enhancement in the footpoint area is only ten times larger than the value in the region of the loop axis. However, in the solar atmosphere the density ratio between the photosphere and the loop top is of the order of 10^{5-6}. Therefore, it is expected to have a higher current-density in actual practice than the present results are indicating, even if the actual rotational velocity is smaller.

Figure 22 shows the spatial distribution of the return-current J_z. As was shown in Figure 21, a direct shell-structured current can be produced, of which the spatial distribution is shown in the figure. The return current has a broad spatial distribution outside the shell containing the direct current. Therefore the current-density of the return-current is weaker than that of the direct current. The direct current tends to pinch near the center of the loop, which, therefore, results in a strong current-loop with a shell structure.

During the propagation of the direct current away from the loop footpoints, a density enhancement appears, to a value a few times larger than the background density. The enhancement also propagates along the loop, with the direct current. This density enhancement may be observed as a travelling soft X-ray enhancement, a feature that is really observed in the initial phase of the small limb flare of 2 August, 1993 (de Jager *et al.*, 1995).

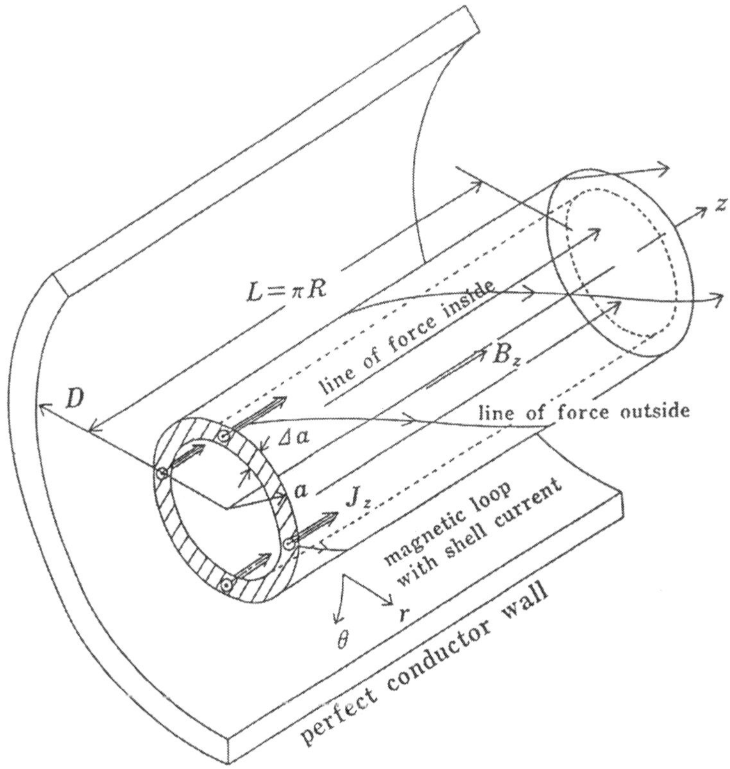

Figure 23. Cylindrical model of a magnetic loop with a shell current. The magnetic lines of force inside the current shell are straight. The lines of force outside the current shell are helical. The conducting wall defines the boundary conditions 'at infinity' (Koide *et al.*, 1994.)

3.3. STABILITY OF LOOPS WITH A SHELL-CURRENT

In the previous section it has been shown that a current-loop can be generated by twisting motions at the loop footpoints. It consists of a shell-current and a diffuse return-current. Therefore it is important to investigate the MHD stability of a loop, containing a shell-current, both with and without a return-current. Previous studies (Mikic *et al.*, 1990; Velli *et al.*, 1990; Strauss, 1990; Zaidman *et al.*, 1991) treated the MHD stability of a loop in which the current occupies the whole region of the loop. Foote and Craig (1990) investigated the ideal MHD stability of a shell-current. On the other hand, Koide and Sakai (1994) investigated the resistive MHD instability of a loop with a shell-current, and showed that such a structure is unstable against tearing instability and breaks into many small current-filaments.

We present here some results on the tearing instability of a loop with a shell-current structure, which were obtained from a 3-D resistive MHD simulation (Koide *et al.*, 1994). Figure 23 shows a cylindrical magnetic loop with length L, radius a, current-shell thickness Δa and current-density J_z. A boundary condition

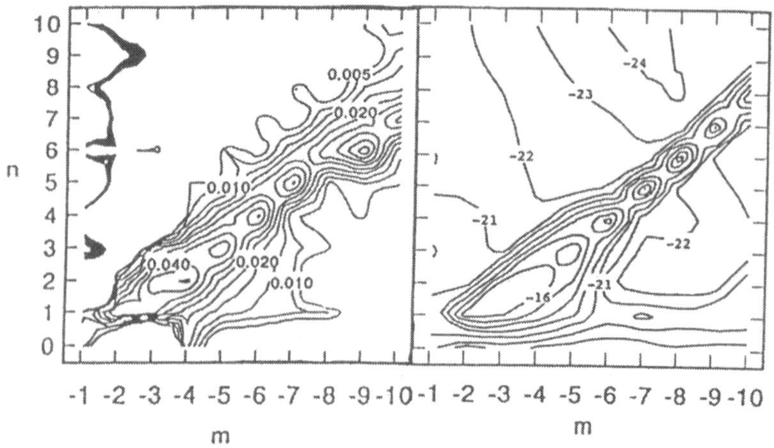

Figure 24. Cylindrical loop with a shell-current; cf., Figure 23. Growth rate (*left*) and magnetic energy (*right*) for the case of a thick shell, a long and thin loop, a low current-density (case (e) – see the text). Here, m and n are wave numbers of Fourier modes, as described in the text. (Koide *et al.*, 1994.)

is imposed by surrounding the region of calculations by a rigid and electrically perfectly conducting wall with radius D. The distance of the wall is taken $D/a = 4$, in order to avoid the disturbing influence from the wall. The outside magnetic lines of force are screwed up with a characteristic distance $2\pi R$; for the aspect ratio A of the loop we take $A = R/a = 2$. For four important parameters, the radius a of the loop, the thickness Δa of the shell current, the length L of the loop, and the current density J_z, the resistive MHD instability of the loop has been investigated:

(a) For a situation defined as a thin shell-current, a long and large loop with a high current-density: $a/\Delta a = 5, A = 4, a = 0.5$, and $J_{z0} = 5$.

(b) For a thick shell-current, a long and thin loop with a high current-density: $a/\Delta a = 2.5, A = 8, a = 0.25$, and $J_{z0} = 5$.

(c) For a thin shell-current, a long and thin loop with a high current-density: $a/\Delta a = 5, A = 8, a = 0.25$, and $J_{z0} = 5$.

(d) For a thick shell-current, a short and thin loop with a high current-density: $a/\Delta a = 2.5, A = 2, a = 0.25$, and $J_{z0} = 5$.

(e) For a thick shell-current, a long and thin loop with a low current-density: $a/\Delta a = 2.5, A = 8, a = 0.25$, and $J_{z0} = 2.5$.

(Note that the adjectives like thick, short, thin, high, ... are only used in a relative sense.)

The growth rate of the instability and the number of current-filaments caused by the instability were found to be determined by the current-density of the loop, rather than by the thickness of the shell or the length of the loop. In the high-current case many current-filaments appear rapidly, while in the low-current case a few current-filaments appear only slowly. Figure 24 presents the growth-rate and the magnetic energy for case (e). For the same case, Figure 25 shows the spatial

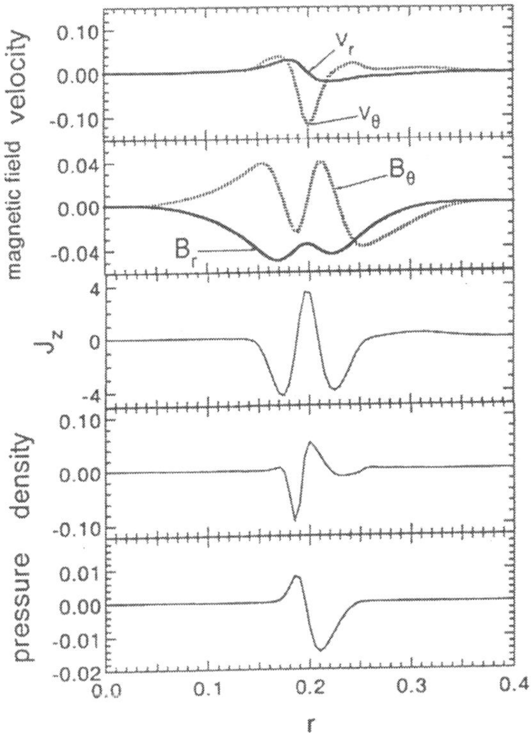

Figure 25. Cylindrical loop with a shell-current; cf., Figure 23. Eigenfunctions of the velocity, magnetic field, current-density, plasma density and pressure for the Fourier modes with wave numbers $m = -4$, $n = 2$, for case (e) (see the text and Figure 24), this being the low current-density case. (Koide *et al.*, 1994.)

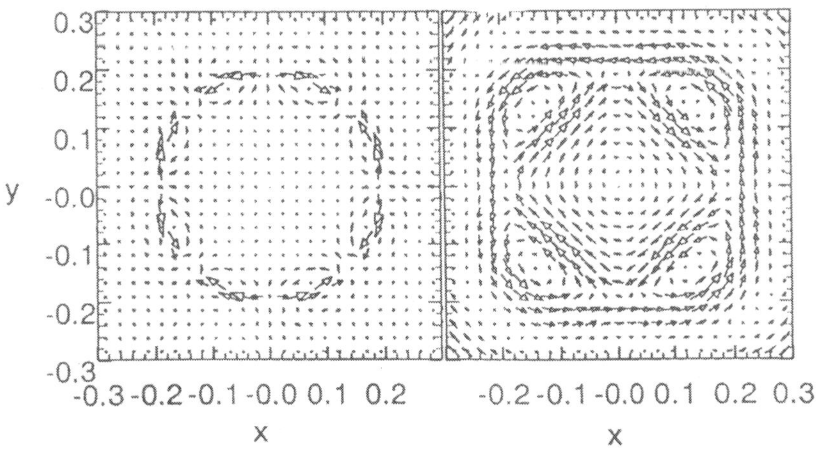

Figure 26. Cylindrical loop with a shell-current; cf., Figure 23. *Left*: full pattern of plasma flow for the Fourier modes $m = -4$, $n = 2$ across the magnetic loop, in the low current-density case (e); see the text, and Figures 24 and 25. *Right*: full pattern of the magnetic field in helical coordinates for wave numbers of modes $m = -4$ and $n = 2$ in case (e). (Koide *et al.*, 1994.)

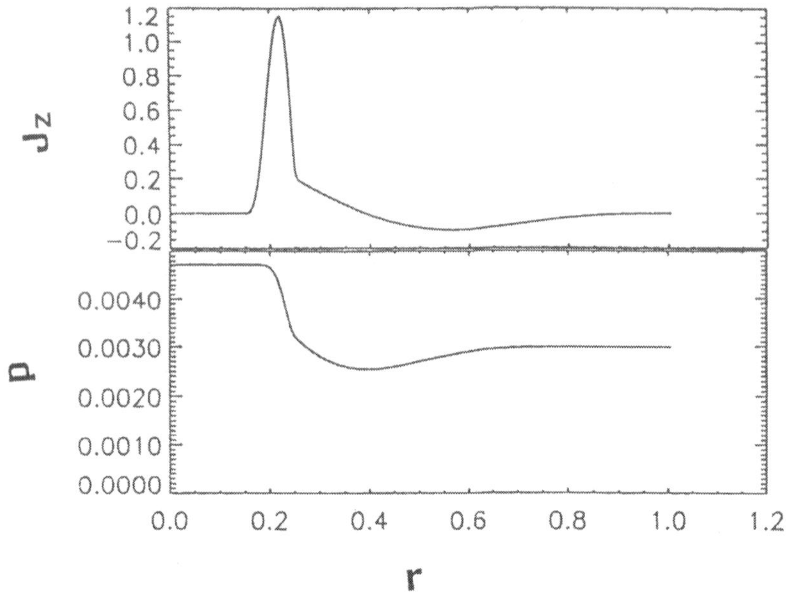

Figure 27. Stability of a loop with a closure current: initial distributions of current J_s (*above*) and pressure P (*below*) in the case T = const. (Koide *et al.*, 1994.)

distributions of the velocity, magnetic field, current density, density, and pressure for the $m = -4$ and $n = 2$ modes. Here, m is the poloidal wave number and n is the wave number in the z direction. Figure 26 gives, again for the same case (e), the flow pattern and the magnetic fields in helical coordinates. From the study of these structures of plasma flow and magnetic field it is concluded that the tearing instability causes magnetic reconnection and that the shell-current breaks into four current-filaments. The threshold of the tearing instability is for $1.4 < q < 2.5$, where $q = rB_z/RB_\theta$ is called the *safety factor*. The threshold is lower than that for a kink instability. This implies that a tearing instability happens easier than a kink instability (cf., also section 5.1 and subsequent sections).

3.4. Nonlinear Dynamics of Many Current-Filaments

In the previous section we showed that many current-filaments may originate in a loop due to the tearing instability which occurs in a loop with a shell-current structure. As discussed in Section 3.1, footpoint rotation generates a loop with a shell-current, associated with a closure current. Koide and Nishikawa (1994) investigated the stability of such a loop with a closure current. Figure 27 shows the initial axial distribution of current and pressure used in their simulation.

Figure 28 shows that interchangeable modes appear, in contrast with the tearing modes that originate in the case of the shell-current loop discussed in the previous section. The figure presents the flow and magnetic field patterns associated with

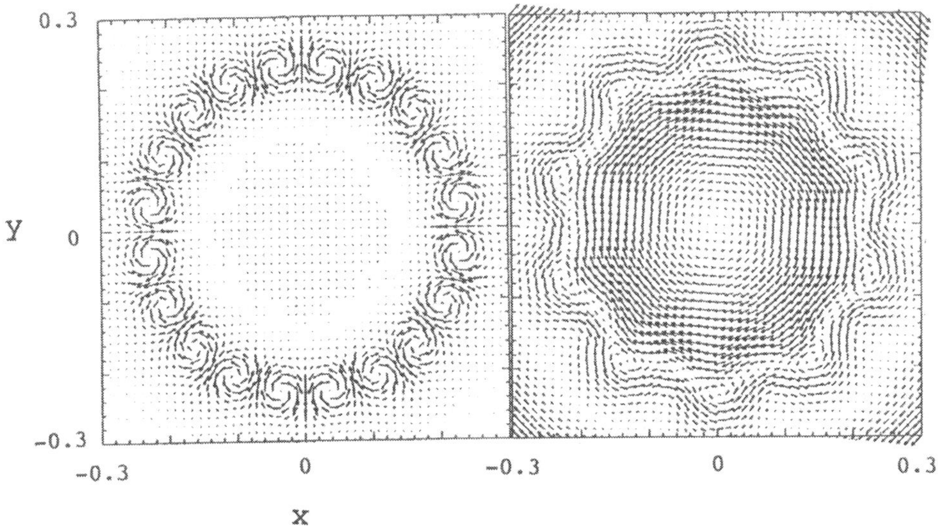

Figure 28. Stability of a loop with a closure current. *Left*: flow patterns, and *right*: magnetic fields, originating in a loop with a closure current, with initial conditions as given in Figure 27. The plots are through a cross section and show the appearance of interchange modes. The diagrams are for modes $m = -10$ and $n = 3$, at $t = 35\tau_A$. (Koide *et al.*, 1994.)

their interchange instability. In this simulation the temperature is initially assumed constant in the radial direction, while the density varies as required by the equilibrium conditions. The magnetic Reynolds number used in the simulation is $S = 10^4$; the plasma beta at the center of the loop is taken 0.005. The flow and magnetic field patterns presented in the above figures correspond to the mode with $m = -10$ and $n = 3$ at the time $t = 35\tau_A$. The growth rate is about $\gamma\tau_A = 0.4$.

Figure 29, which shows the current through a cross-section, demonstrates that both the direct current-shell and the return current-shell tend to split into many filaments. In addition, it is found that the filaments associated with the direct current (corresponding with the white regions in the figure), show a weaving structure.

Koide and Nishikawa (1994) also derived the nonlinear evolution for the case of a shell-current loop with a closure current, where the density is initially taken to be constant, while the temperature varies according to the equilibrium conditions. Figure 30 presents the assumed initial current- and pressure distributions. In this case tearing modes appear; the dominant mode numbers (m, n) are $(-2, 1)$, $(-4, 2)$, and $(-5, 3)$. The typical growth rate is about $\gamma\tau_A = 0.1$. The magnetic Reynolds number used in this simulation is $S = 10^4$, the plasma beta at the axis of the loop is about 0.025. Figure 31 show the time evolution of the current J_z at different times. After filamentation of the linear stage of the tearing instability, adjacent filaments tend to coalesce, as seen in the figure.

The main result of these studies is that they show that an initially single loop with a shell-current does not remain homogeneous but develops many filaments

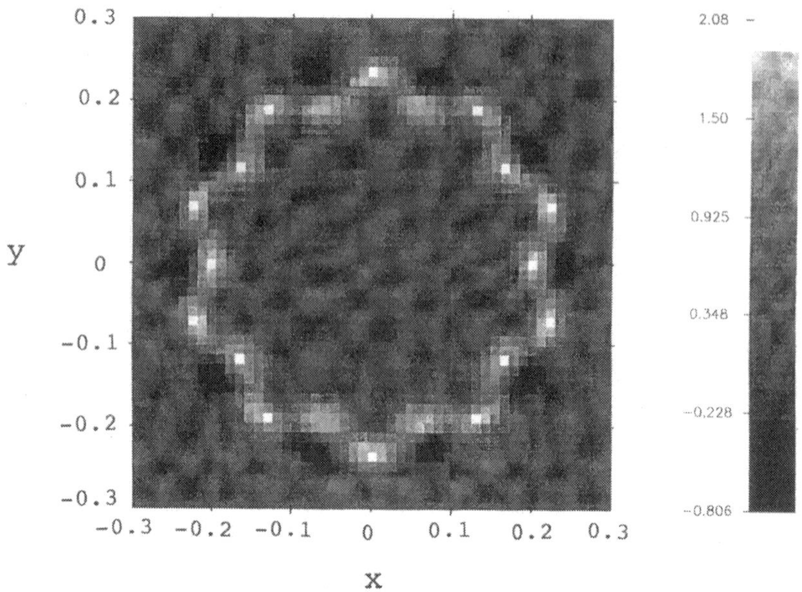

Figure 29. A loop with a shell-current. The distribution of current J_z, associated with the interchange modes; cf., Figures 27 and 28. (Koide *et al.*, 1994.)

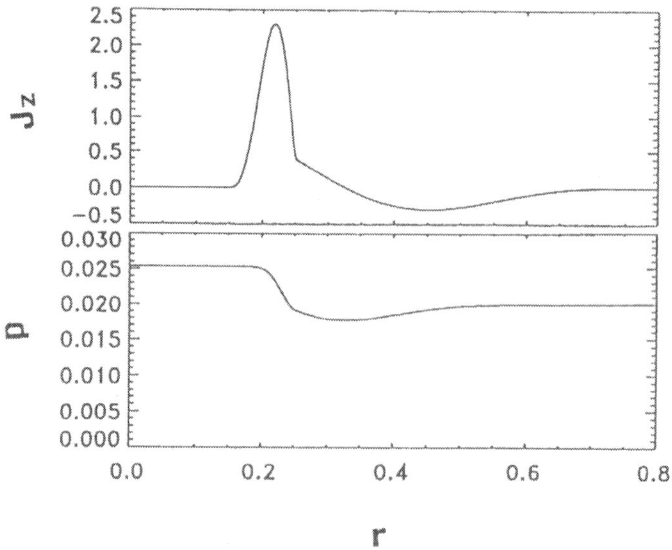

Figure 30. A loop with a shell-current. Initial distributions of the z-component of the current J_z, and the pressure P, in the case $r =$ const. (Koide *et al.*, 1994.)

inside the current-loop. This seems to be essential for understanding the process of coronal loop heating as well as the properties of single-loop flares. Therefore, it is worth while to investigate the plasma heating process in some detail. We mention

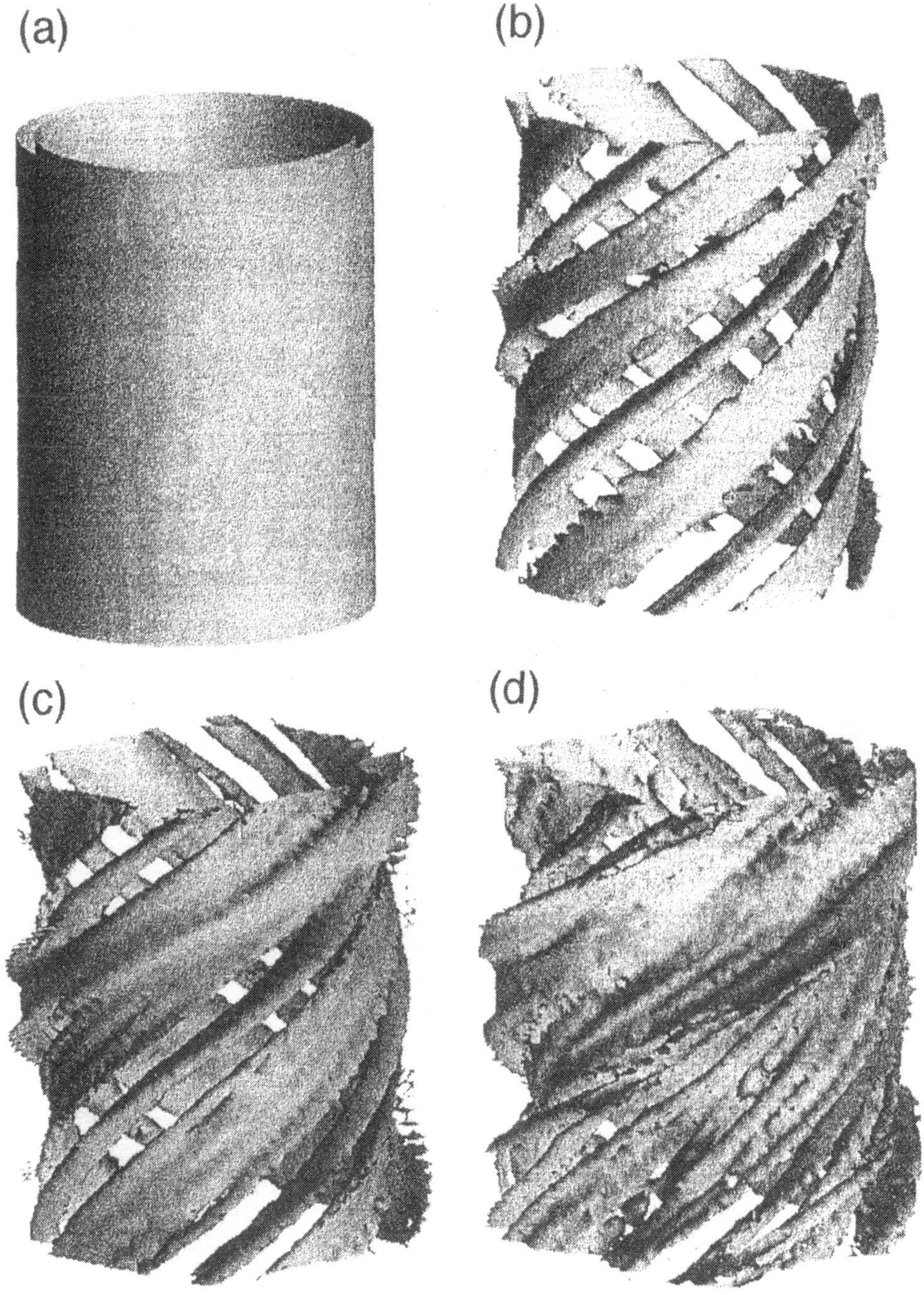

Figure 31. Nonlinear development of the current J_z in a fluxtube with a shell-current, associated with the initial conditions assumed in Figure 30: (a) isosurface $J_z = 2$ at $t = 50\tau_A$, (b)–(d) isosurfaces $J_z = 2.4$ at $t = 60$, 70, and $80\tau_A$, respectively. (Koide *et al.*, 1994.)

here that shell-currents seem to exist in loop-prominences. Because of the observed radiative cooling near the central region of the loop, the current in the prominence may be localized within its shell structure, in the way as described in the previous section. Therefore, in the eruptive phase of loop prominences many thin filaments (fluxthreads) may occur inside the loop, similar to the filaments shown in Figure 31. We conclude that the study of the nonlinear dynamics of a loop with a shell-current structure is important both for understanding the current-generation phase as well as for the current-decay phase, which is due to radiative cooling. The latter phase may correspond to the later phases of post-loop flares, as described in section 2.3.4.

4. Coalescence of Two Current-Loops

4.1. Partial and Complete Reconnection

Sakai and Koide (1992b) proposed a classification scheme for the cases of magnetic reconnection during coalescence of two current-loops. This classification may be important for understanding the physical processes, both in solar flares and in coronal loop heating in solar active regions. We consider first a coronal loop extending along the z-direction with a current I_z, whose footpoints are assumed to be situated in an active region. The magnetic-field component B_θ is determined from the current I_z . We assume that the direction of the magnetic field component B_z along the loop is determined by the magnetic polarities at the footpoints. We thus find that the current-loop has a positive rotational angle when the magnetic field vector component B_z and the current-vector component I_z have the same directions, while the current-loop has a negative rotational angle when B_z and I_z have opposite directions.

We next consider the interaction of the two current-loops, with the aim to examine what kinds of magnetic reconnection can occur near the interaction region of the two loops. There are two types of interaction, depending on whether the magnetic field components, B_z of the two loops are parallel or anti-parallel. Furthermore, there are two cases of interaction, depending on whether the current-directions of the loops are parallel or anti-parallel. The current-direction along the loops is determined from the (complicated) motions of the footpoints. As shown in Section 3, a loop with a current-shell structure associated with a return current can become unstable and can split into many filaments. In the nonlinear stage of two-loops interaction in which many filaments interact, the type of classification of the reconnection is therefore also important.

The combination of the interaction classifications described above, gives us six different kinds of current-loop coalescence processes. The first case is shown in Figure 32, in which the left-hand part shows the state of the current-loops before magnetic reconnection, while the right-hand part shows it after reconnection. When the two current-loops are completely parallel, as shown in Figure 32, only the

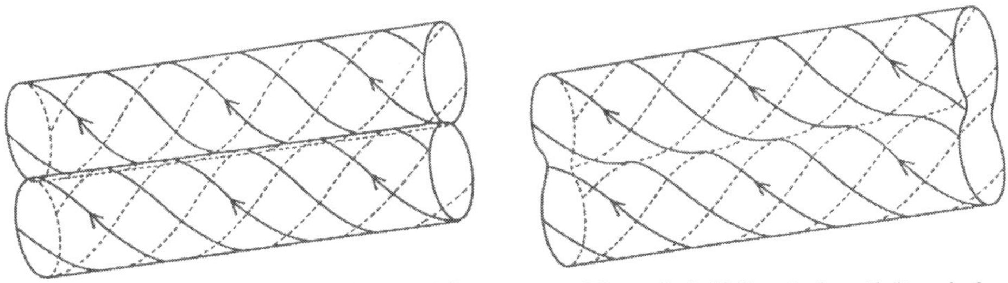

Figure 32. Parallel loops with parallel field-line systems. Magnetic field lines before (*left*) and after (*right*) partial reconnection. (Sakai and Koide, 1992b. From *Solar Phys.*)

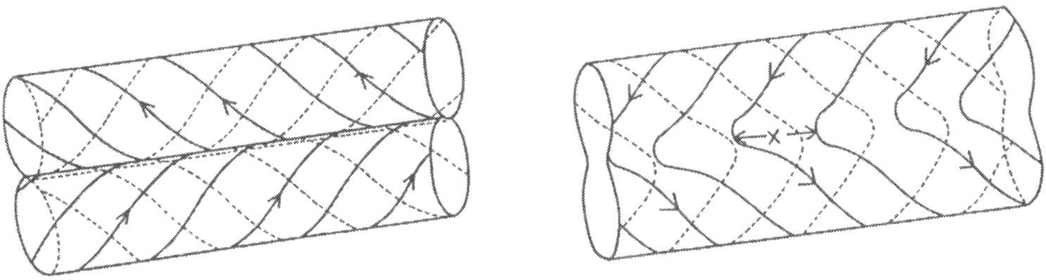

Figure 33. Parallel loops with oppositely directed field-line systems. Magnetic field lines before (*left*) and after (*right*) complete reconnection. (Sakai and Koide, 1992b. From *Solar Phys.*)

magnetic field component B_θ can dissipate through reconnection. We call this *partial reconnection*. However, if the two loops collide with each other under some angle, as happens with X-type coalescence (Sakai and de Jager, 1991) both magnetic components B_θ and B_z can be dissipated through reconnection. We call this *complete reconnection*; cf., Figure 34.

Cheng *et al.* (1994) published evidences both for cases of partial reconnection as well as for complete reconnection, as observed in soft X-ray *Yohkoh* images. A few evidences (Takahashi *et al.*, 1995; Inda-Koide *et al.*, 1995; de Jager et al., 1995) of partial reconnection were also found in small flares observed by *Yohkoh*. These, as well as some other selected flare observations are discussed in Section 6 of this review.

The amount of energy released in complete reconnection is larger than in the case of partial reconnection. Also, the rate of reconnection is larger in the case of complete reconnection than in that of partial reconnection, as appears from laboratory experiments (Yamada *et al.*, 1990; Ono *et al.*, 1993). The rate of reconnection depends on the colliding velocities of the two loops (Yamada *et al.*, 1990) as well as on the plasma β ratio (Sakai and de Jager, 1991). The reconnection can become explosive (Sakai and Ohsawa, 1987; Tajima and Sakai, 1989b) and will proceed very rapidly when the plasma β ratio is much smaller than unity, while the reconnection rate becomes slow when the plasma β ratio approaches unity. This is the case even when the two loops are forced to collide.

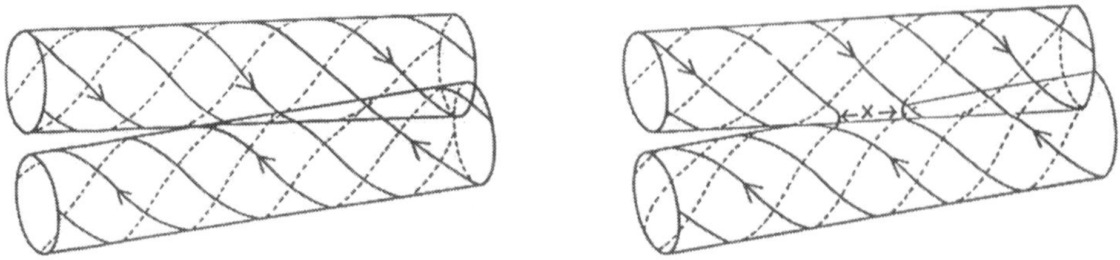

Figure 34. Oppositely directed field-line systems in current-loops, meeting under a certain angle. Magnetic field lines before (*left*), and after reconnection (*right*) for X-type coalescence. (Sakai and Koide, 1992b. From *Solar Phys.*)

The other important parameter controlling the reconnection rate in the case of partial reconnection is the ratio between B_z and B_θ. If $B_z > B_\theta$, there occurs a tilting motion near the reconnection point, which can produce plasma jets that can be observed as a coronal explosion (Sakai and de Jager, 1989a; Sakai, 1989; Chargeishvili *et al.*, 1993). This is the reason why plasma jets can occur with confined flares as well as with eruptive/dynamic flares (cf., Section 2.3.2).

Fushiki and Sakai (1995a) investigated the two-loops coalescence process by using a 3-D resistive MHD code, in which particular attention was given to the processes of partial and complete reconnection. They found that the formation of plasma jets as well as of fast magnetosonic waves can occur during the process of coalescence of two current-loops. They took for the initial current-loops a pair of force-free current-loops as introduced by Gold and Hoyle (1960). The magnetic fields are given by

$$B_\theta = \frac{B_m r/a}{1+(r/a)^2}, \qquad B_z = \frac{B_m}{1+(r/a)^2},$$

where a is the radius of the loop and r the distance from the axis. For the case of partial reconnection, B_m has the same sign for both loops, while for complete reconnection opposite signs are assigned to the two loops. The initial density was taken to be constant (unity) and velocity perturbations were imposed in order to allow for a local collision. The time history of the density of the two loops is shown for the case of complete reconnection in Figures 35 and 36. The two current-loops appear to collide locally and the density near the coalescence region is enhanced. Figure 37 shows the isosurfaces of the total current-intensity and of the total magnetic field strength at $t = 2.5\tau_A$. Figure 38 presents the magnetic field lines at $t = 2.5\tau_A$. The figure demonstrates that complete magnetic reconnection can occur in a plane oblique to the z-axis. This complete reconnection ultimately drives the two plasma flows away from the region of reconnection, as seen in Figure 39. During the process of two-loops coalescence two plasmoids are formed, as seen in Figure 40.

On the other hand, in the case of partial reconnection, where the B_z components of the two loops have the same sign, only the poloidal components of the magnetic

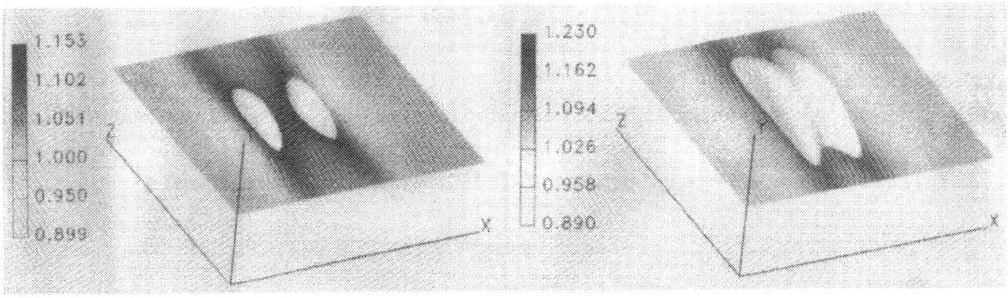

Figure 35. Two-loops complete coalescence. The figure combines the density distributions (gray scales) at the $y = 3.14$ level and isosurfaces ($\rho = 1.1$; white areas) of densities. The diagram shows the development at $t = 0.5\tau_A$ (*left*) and $t = 1\tau_A$ (*right*). (Fushiki and Sakai, 1995a. From *Solar Phys.*)

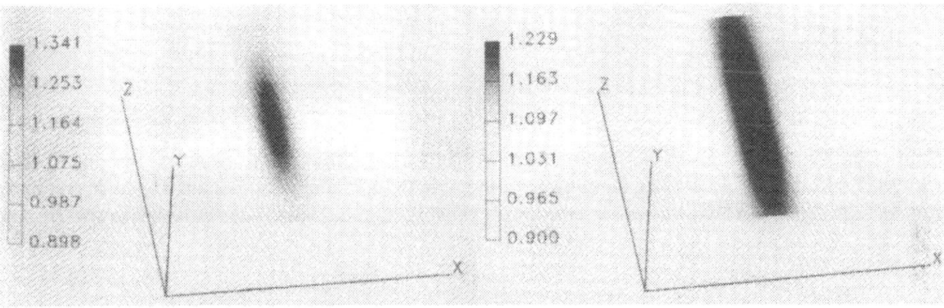

Figure 36. Two-loops coalescence. As Figure 35, but for $t = 1.5\tau_A$ (*left*) and $t = 2\tau_A$ (*right*).

fields can reconnect, as seen in Figure 41. In that case the plane of reconnection is perpendicular to the z-axis. It was also found (Sakai and Fushiki, 1995), that after the event of local loop coalescence with partial reconnection, strong fast magnetosonic waves can be produced. The waves appear to propagate away from the coalescence region, cf., Figure 42. For the case of partial reconnection during two-loops coalescence the formation of shocks was also confirmed, by calculations based on MHD simulations (Koide and Sakai, 1994). The situation becomes more

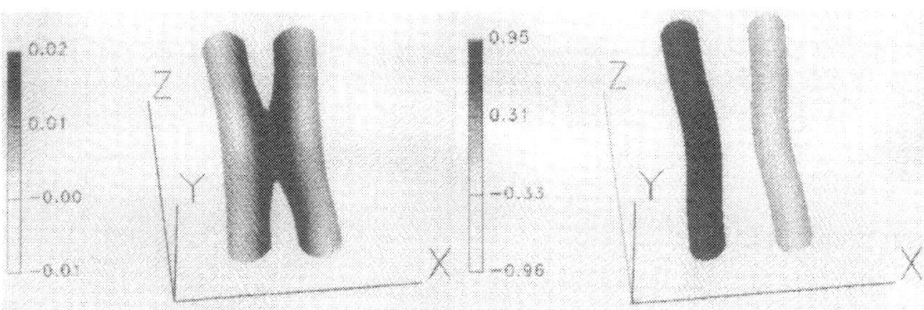

Figure 37. Two-loops complete coalescence at $t = 2.5\tau_A$; cf., Figures 35 and 36. Isosurface ($|\mathbf{J}| = 0.01$) of the total current; the gray-scale is the current J_z (*left*). *Right*: isosurface ($|\mathbf{B}| = 0.712$) of the total intensity of the magnetic fields; the gray scale is the magnetic field B_z. (Fushiki and Sakai, 1995a. From *Solar Phys.*)

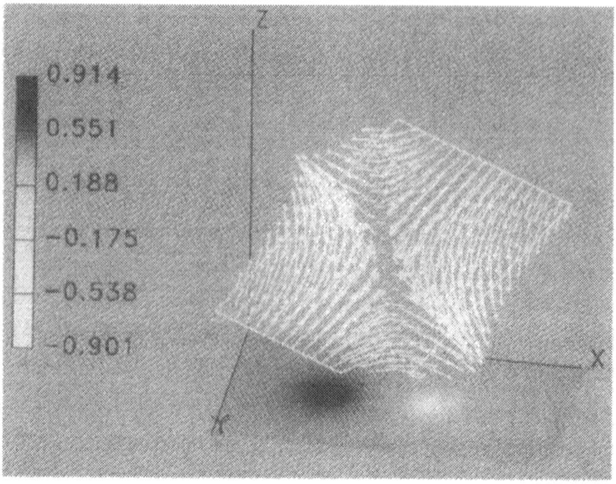

Figure 38. Two-loops complete coalescence. Magnetic field lines in the plane of complete reconnection ($t = 2.5\tau_A$). The gray-scale gives the magnetic field strength component B_z, as in the lower frame of the Figure, which shows the two loops. (Fushiki and Sakai, 1995a. From *Solar Phys.*)

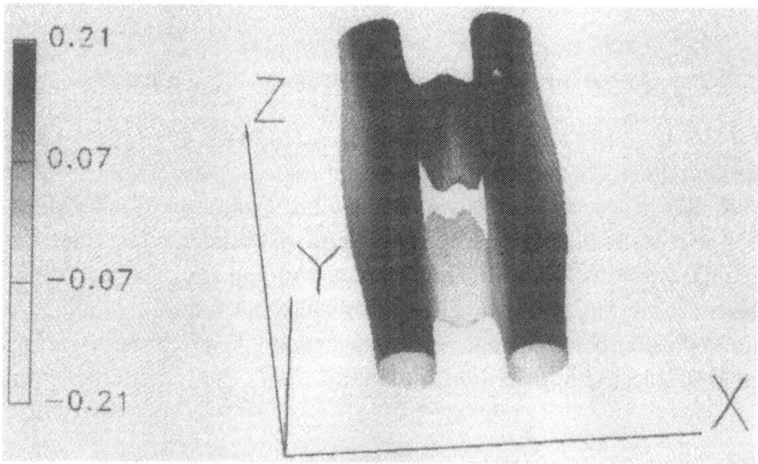

Figure 39. Two-loops complete coalescence. Velocity field in the case of complete reconnection (cf., Figuress 35–38). The figure shows the isosurface ($|\mathbf{v}| = 0.22$) of the total intensity; the gray-scale gives the velocity component V_z ($t = 2.5\tau_A$). (Fushiki and Sakai, 1995a. From *Solar Phys.*)

favorable for the formation of shock-waves as well as for that of plasma jets when the plasma beta is small.

4.2. Plasmoid Formation during Two-Loops Coalescence

By using a 3-D resistive MHD code Fushiki and Sakai (1995a) showed how plasmoids may originate during two-loops current-coalescence. Plasmoids can

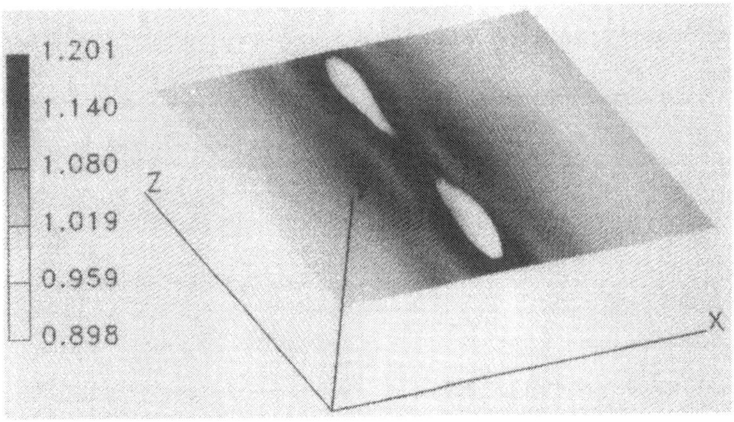

Figure 40. Two-loops coalescence. Plasmoid formation at $t = 2.5\tau_{\rm A}$ (cf., Figures 35–38). Isosurface ($\rho = 1.19$) of density; the gray-scale is the density distribution at the level $y = 3.14$, normalized to the starting value. (Fushiki and Sakai, 1995a. From *Solar Phys.*)

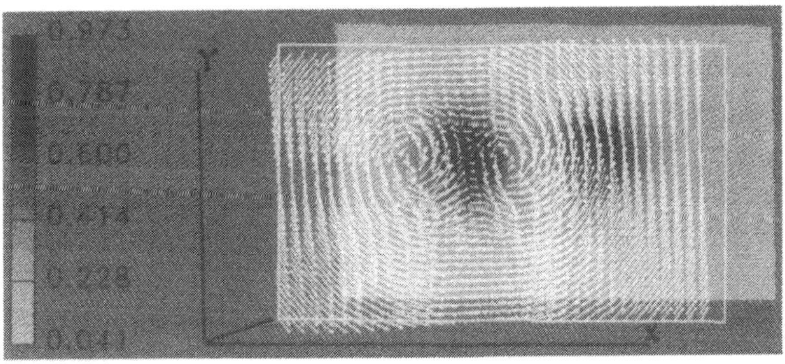

Figure 41. Case of partial reconnection. The B_z components of the two loops have the same sign: only the poloidal components reconnect. Magnetic field lines in the plane of partial reconnection at $t = 2.5\tau_{\rm A}$; the gray-scale is the magnetic field B_z. (Fushiki and Sakai, 1995a. From *Solar Phys.*)

originate when two current-loops collide in the case of complete reconnection as described in the previous section, and when furthermore the two loops collide locally in at least two different points. Multiple local collision between two loops may occur when the two current-loops, or one of them, become kink-unstable and collide.

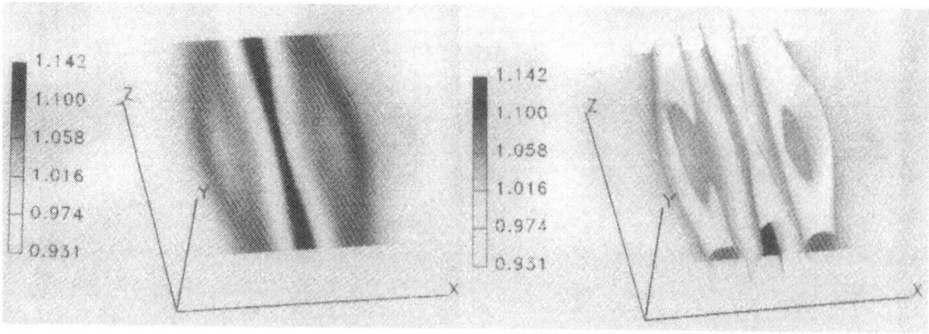

Figure 42. Partial reconnection (cf., Figure 41). Density distribution (*left*) at the $y = 3.14$ level at $t = 4\tau_A$. The isosurface ($\rho = 1.04$) (*right*) at $t = 4\tau_A$ shows the magnetosonic wave emission (wave amplitude $\simeq 0.1$) from the coalescence region. (Fushiki and Sakai, 1995a. From *Solar Phys.*)

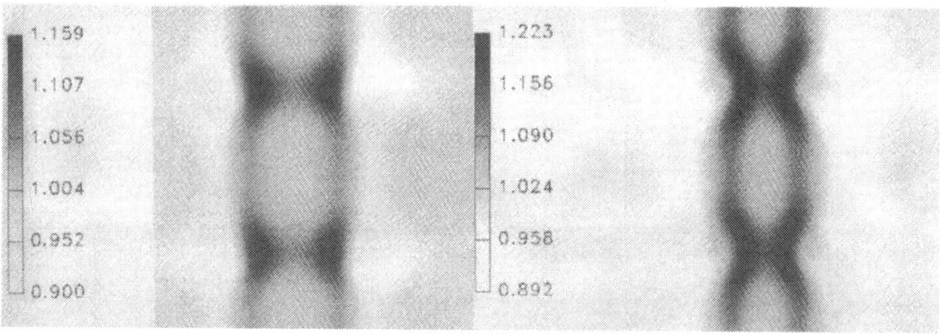

Figure 43. Plasmoid formation in loops that collide locally in two different points. Density distributions at the $y = 3.14$ level at $t = 0.5\tau_A$ (*left*) and $t = 1\tau_A$ (*right*) for complete reconnection with $\beta = 0.5$. (Fushiki and Sakai, 1995a. From *Solar Phys.*)

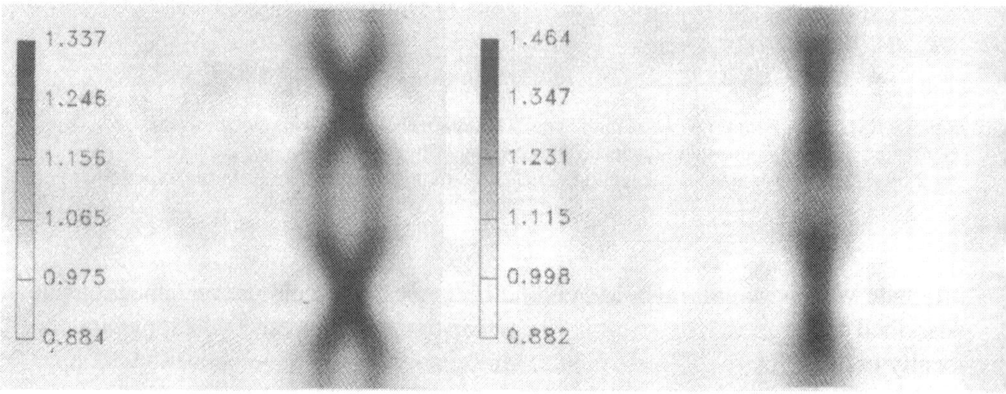

Figure 44. Plasmoid formation. As Figure 43 but for $t = 1.5\tau_A$ (*left*) and $t = 2\tau_A$ (*right*).

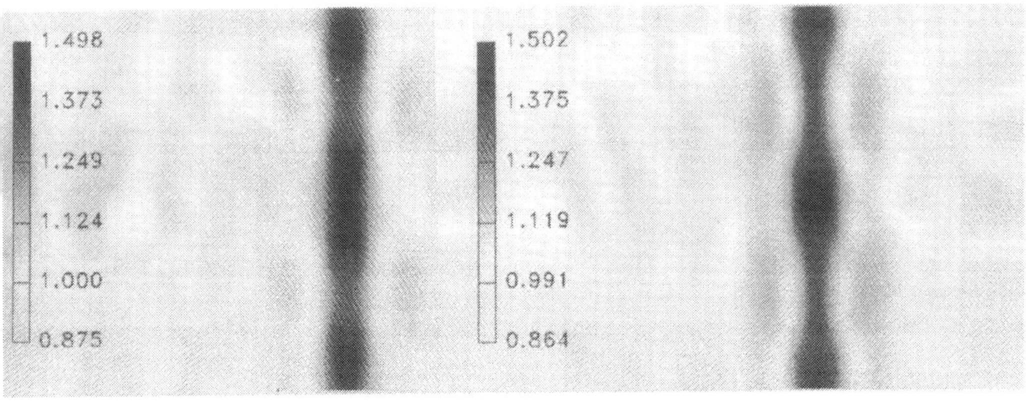

Figure 45. Plasmoid formation. As Figure 43, but for $t = 2.5\tau_A$ (*left*) and $t = 3\tau_A$ (*right*).

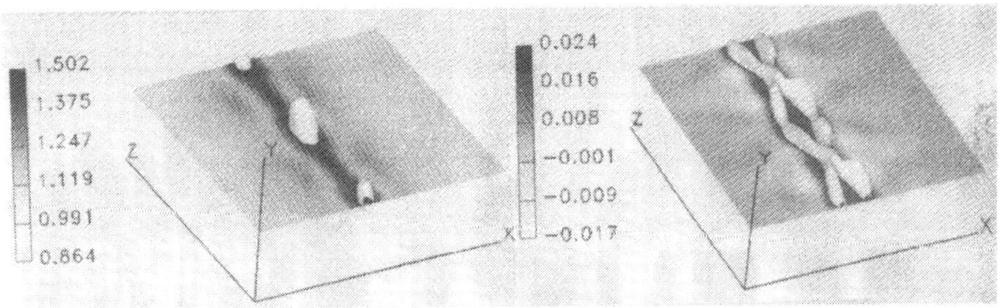

Figure 46. Plasmoid merging in the case of two loops that collide in two different points (cf., Figures 43–45). Isosurface ($\rho = 1.407$) of the density (*left*) and isosurface ($J_y = 0.0146$) of the current (*right*) at $t = 3\tau_A$. (Fushiki and Sakai, 1995a. From *Solar Phys.*)

Figures 43–45 show the time history of the density of two loops that collide locally at two different points. The initial current-configuration is taken identical to that of the force-free field described in the previous section. At the two points of collision the complete reconnection can yield oppositely directed plasma jets. The plasma jets, flowing away from the two regions of complete reconnection, may collide and merge into one plasmoid, as shown in Figure 46.

The plasmoid can move into one direction if the two plasma flows are not symmetric. This may be due to a variety of causes (for example, different current-intensities, different colliding velocities, and different loop radii) with reference to the reconnection at the two points of coalescence. Such asymmetric conditions may lead to rotational motions in the plasmoid. This plasmoid formation process may have been observed by *Yohkoh* (Tsuneta, 1993) in the flare of 2 December, 1991.

To demonstrate an aspect of the case of current-loops with non-symmetric conditions at the points of reconnection, Fushiki and Sakai (1995b) found that helical plasma flows originate during collisions of two current-loops carrying angular

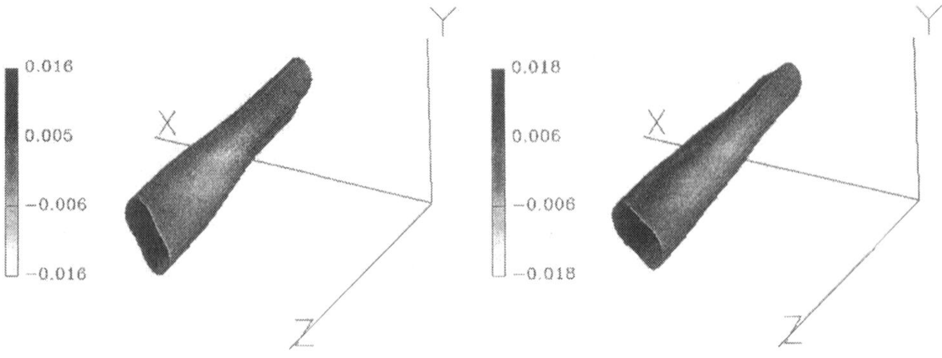

Figure 47. Helical plasma flow formation in colliding loops carrying angular momentum. Isosurface ($|\mathbf{J}| = 0.019$) of the total current intensity at $t = 0.5\tau_A$ (*left*) and at $t = 1\tau_A$ (*right*) in the case of partial reconnection. The gray-scale is the current J_z. (Fushiki and Sakai, 1995b. From *Solar Phys.*)

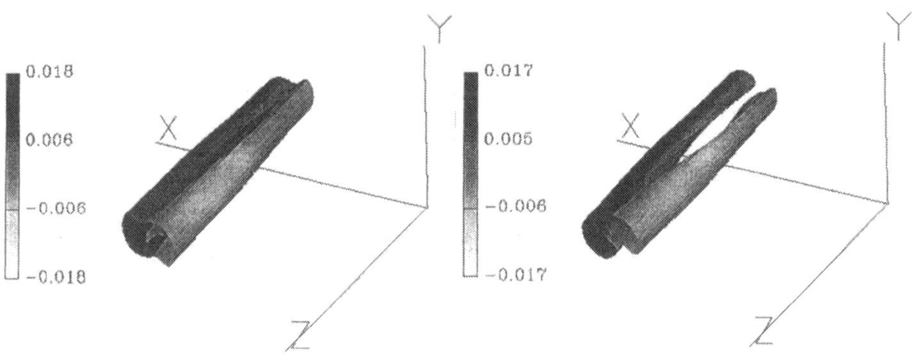

Figure 48. Helical plasma flow formation. As Figure 47, but for $t = 1.5\tau_A$ (*left*) and $2\tau_A$ (*right*).

momentum. Figures 47 and 48 show the time development of the isosurfaces of the total current intensity during the process of two-loops collisions.

Figures 49 and 50 present the shape and intensity of the helical plasma flow produced during collisions of two current-loops. The process of formation of helical plasma flows during two-loops collisions may be related to some surges and X-ray jets as observed by *Yohkoh* (Yokoyama and Shibata, 1994; Shibata *et al.*, 1994a, b).

4.3. Magnetic Reconnection during Two-Loops Coalescence

In this section we review the process of magnetic reconnection during coalescence of two current-loops. Previous studies, till 1987, were summarized in a review paper by Sakai and Ohsawa (1987). Magnetic reconnection in the MHD approximation is summarized by Biskamp in his book and his review paper (Biskamp, 1993, Section 6; Biskamp, 1994). As pointed out there, it is important to note that the classical steady reconnection model by Petschek (1964) is not a self-consistent

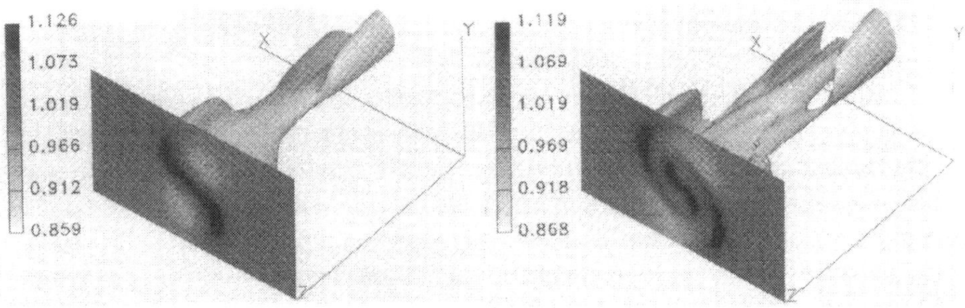

Figure 49. Helical plasma flow formation (cf., Figure 47). Isosurfaces ($\rho = 1.06$) of the density, for $t = 0.5\tau_A$ (*left*) and $t = 1\tau_A$ (*right*). (Fushiki and Sakai, 1995b. From *Solar Phys.*)

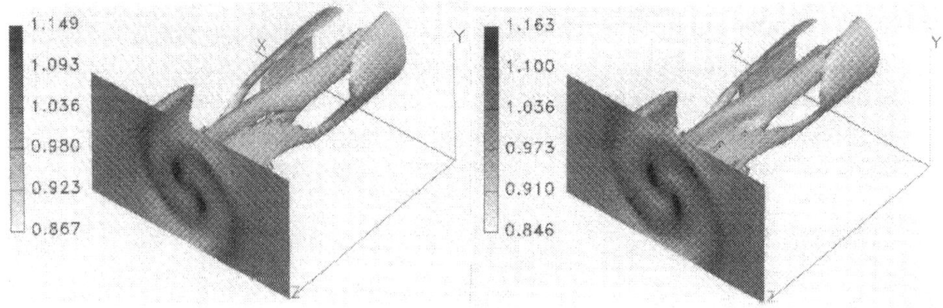

Figure 50. Helical plasma flow formation. As Figure 49, but for $t = 1.5\tau_A$ (*left*) and $2\tau_A$ (*right*).

reconnection model in the limit of small resistivity. The crucial deficiency of the reconnection models of the Petschek type is the neglect or inappropriate treatment of the diffusion region. It is also important to consider the tearing mode stability (Bulanov *et al.*, 1979; Phan and Sonnerup, 1991) in a current-sheet with finite size and the related plasma flow and its nonlinear behavior (Biskamp, 1986) in a dynamic Sweet–Parker sheet for sufficient small resistivity. Magnetic reconnection in collision-less plasmas is discussed in our Section 5.

As pointed out by Sakai and Ohsawa (1987), the magnetic reconnection processes can be classified in different ways. We have *steady* or *non-steady* reconnection, depending on whether the reconnecting flux develops proportionally to the time t or not. In addition, reconnection is *spontaneous* or *driven*, depending on the boundary conditions. The reconnection process during current-loop coalescence is therefore a non-steady and driven process. The case of driven reconnection during current-loop coalescence can be further classified into three types; (1) *steady*, (2) *fast*, and (3) *explosive* magnetic reconnection. These three types have been studied in 2D-MHD simulations (Bhattacharjee *et al.*, 1983; Brunel *et al.*, 1982; Tajima and Sakai, 1989b). The fastest, *explosive reconnection*, is characterized by the fact that the reconnecting flux is independent of the resistivity and proportional to $(t_0 - t)^{-4/3}$, where t_0 is called the explosion time (Sakai and Ohsawa, 1987). The case of explosive reconnection was also found in a 2-D particle simulation and the explosiveness of the physical quantities derived from theory has been confirmed by simulation results (Tajima and Sakai, 1989b).

Shibata *et al.* (1989, 1992) have carried out extensive 2-D MHD simulations, in order to study the magnetic reconnection between an emerging flux and an overlying coronal magnetic field, while including the effect of gravity. The reconnection starts when most of the chromospheric mass in the current-sheet between the emerging flux and the coronal field has drained away, down along the loop, due to gravity. They also showed that multiple magnetic islands are created within the sheet; these islands confine a cool, dense chromospheric plasma. They coalesce dynamically with each other and are thereupon ejected along the sheet. Matsumoto *et al.* (1993) studied the nonlinear evolution of an emerging magnetic flux tube in the solar atmosphere by 3-D MHD simulations. They showed that the magnetic flux sheet or tube is unstable against the undular mode of the magnetic buoyancy instability. During the ascendance of loops the bundle of flux tubes develops into dense gas filaments pinched between magnetic loops. The interchange modes produce a fine fiber flux structure perpendicular to the magnetic field direction in the linear stage, while the undular modes determine the overall buoyant loop structure.

4.4. Unified Treatment of I-, Y-, and X-Types of Two-Loops Collisions

The high-resolution soft X-ray images of the corona obtained by the *Yohkoh* mission reveal the fine structure of coronal active regions. Thus, it becomes possible to obtain a more detailed and more precise view of the coronal loop configurations. Using *Yohkoh*'s data, Shimizu *et al.* (1992) have found that from 144 transient brightening events, 61 are single loop events, 69 are multiple loop events, and 14 are point-like events. Furthermore, in the multiple loop events they found that 30% have the X-type configuration (the geometry is shown in Figure 3), 57% are of the Y-type and 13% are of other types. Sakai and Koide (1992) have classified the geometry of magnetic reconnection during current-loop coalescence. They considered all possible mutual positions and current directions of loops for partial or complete reconnection. It appears that under certain conditions the X-, Y- and I-type two-loops configurations can provide both complete or partial reconnection.

Chargeishvili *et al.* (1993) investigated the physical processes that may take place in all cases of X-, Y-, and I-type two-loops interactions. In Figure 51 we present a generalized scheme of two-loops interactions, which describes in a unified way the magnetic field configuration during reconnection in any of the X-, Y- or I-type interactions. Reconnection of the types X, Y, or I correspond in Figure 51 with $L_r/L_z \gg 1$, $= 1$, or $\ll 1$, respectively. Here, L_r and L_z are the axial and radial characteristic lengths of the region of interaction. The reconnection region is considered to be a cylinder with the axis through the RP (Reconnection Point; point where the reconnection takes place) and directed along the line of symmetry of the colliding loops. Cylindrical coordinates (r,ϕ,z) were used. The z-axis is directed

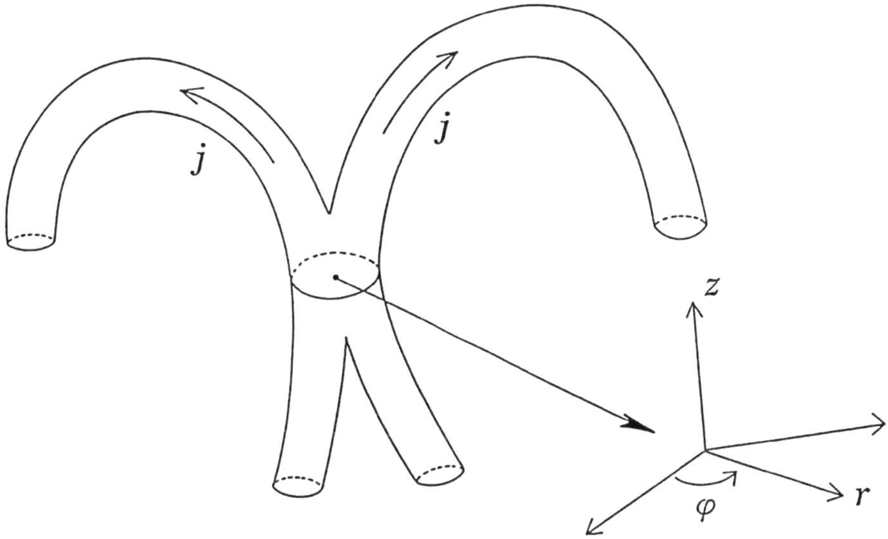

Figure 51. Schematic configuration of the three cases of collisions between two loops in a coordinate system with its origin in the RP (point where reconnection takes place), and the z-axis directed along the line of symmetry of the colliding loops. (Chargeishvili *et al.*, 1993. From *Solar Phys.*)

along the line of symmetry of the colliding loops and the origin of the reference frame coincides with the RP.

When the longitudinal component of the magnetic field is at right angles to the gravity force vector, the atmospheric stratification effect due to gravity is negligible, because the radius of the loop is much smaller than the atmospheric scale height H in the solar radial direction. The authors only consider the atmospheric stratification effect when the gravity force vector is parallel to the longitudinal component of the magnetic field. As will appear from the system of equations governing the plasma behavior and from the numerical simulation results, the stratification has no effect on the temporal evolution of most physical quantities and only a negligible influence on some of them. Hence the results are also valid for the case when the longitudinal component of the magnetic field is perpendicular to the gravity vector.

For treating the physical processes in the three types of two-loops interactions we introduce the following set of standard MHD equations:

– the continuity equation

$$\frac{\partial \rho}{\partial t} + \nabla \cdot (\rho \mathbf{V}) = 0, \tag{1}$$

– the energy equation

$$\frac{\partial P}{\partial t} + (\mathbf{V} \cdot \nabla) P + \gamma P (\nabla \cdot \mathbf{V}) = (\gamma - 1) \frac{\mathbf{J}^2}{\sigma}, \tag{2}$$

– the induction equation

$$\frac{\partial \mathbf{B}}{\partial t} = \nabla \times (\mathbf{V} \times \mathbf{B}) + \frac{c^2}{4\pi\sigma}\Delta\mathbf{B}, \tag{3}$$

$$\nabla \cdot \mathbf{B} = 0,$$

– and the equation of motion

$$\rho\left(\frac{\partial \mathbf{V}}{\partial t} + (\mathbf{V}\cdot\nabla)\mathbf{V}\right) = -\nabla P + \frac{1}{4\pi}(\nabla \times \mathbf{B}) \times \mathbf{B} + \rho\mathbf{g}, \tag{4}$$

where ρ and P are the plasma density and pressure respectively, **V** is the plasma flow velocity, **B** the magnetic field, **g** the free fall acceleration (opposite to the z axis), γ is the adiabatic ratio and σ is the electric conductivity. The right-hand side of the energy equation (2) describes the Ohmic heating, and

$$\mathbf{J} = \frac{c}{4\pi}\nabla \times \mathbf{B}. \tag{5}$$

The authors consider the time evolution of the physical state parameters just after the sudden change that occurred due to magnetic field reconnection, by writing the following equations for the physical quantities, valid in a restricted volume near the RP:

– for the velocity components

$$V_r(t) = \frac{\dot{a}(t)}{a(t)}r \ , \ V_\varphi(t) = v_\varphi(t)\frac{r}{\lambda} \ , \ V_z(t) = V_{z0}(t) + \frac{\dot{b}(t)}{b(t)}z \ , \tag{6}$$

– for the magnetic field components

$$B_r(t) = b_r(t)\frac{r}{\lambda} \ , \ B_\varphi(t) = b_\varphi(t)\frac{r}{\lambda} \ , \ B_z(t) = B_{z0}(t) + b_z(t)\frac{z}{\lambda} \ , \tag{7}$$

where the dot means the time-derivative. The time-dependent values, $a(t)$, $v_\varphi(t)$, $v_{z0}(t)$, $b(t)$, $b_r(t)$, $b_\varphi(t)$, $b_z(t)$, and $B_{z0}(t)$ were self-consistently determined from the MHD equations (1)–(5) as will be shown below. The plasma density is

$$\rho = \frac{1}{a^2 b}, \tag{8}$$

where the density is normalized to ρ_0, which is a constant, equal to the density at the time $t = 0$.

The magnetic field components obey to the following expressions:

$$b_r = \frac{b_{r0}}{a^2 b} \ , \ b_\varphi = \frac{b_{\varphi 0}}{a^2 b} \ , \ b_z = -2\frac{b_{r0}}{a^2 b}, \tag{9}$$

where b_{r0} and $b_{\varphi 0}$ are constants that are also fixed at $t = 0$. For the homogeneous part of the z component of the magnetic field one gets

$$\dot{B}_{z0} = 2B_1 \frac{v_z}{a^2 b} - 2\frac{\dot{a}}{a} B_{z0}, \tag{10}$$

where $B_1 = b_{r0}/b_{\varphi 0}$. It must be mentioned that $B_{z0}(t = 0) = 0$ corresponds to complete reconnection. However, when $B_{z0}(t = 0) \neq 0$, then one deals with partial reconnection, in which the whole poloidal component but only a part of the longitudinal component of the magnetic field reconnects.

The plasma pressure is written as

$$P(t) = P_0(t) - p_H(t)\frac{z}{H} - \frac{p_r(t) r^2 + p_z(t) z^2}{2\lambda^2}, \tag{11}$$

where the second term at the right-hand side describes the stratification of the atmosphere due to gravity and where H is the pressure scale height.

It is then found, from Equation (2), that the time-dependent coefficients in Equation (11) can be written as

$$p_r = \frac{1}{a^{2\gamma+2} b^{\gamma}}, \tag{12}$$

$$p_z = \frac{1}{a^{2(\gamma+1)} b^{\gamma}}, \tag{13}$$

$$\dot{P}_0 = -\gamma P_0 \left(2\frac{\dot{a}}{a} + \frac{\dot{b}}{b} \right) + h p_H v_z + \frac{4(\gamma - 1)}{\beta S} \frac{1}{a^4 b^2}, \tag{14}$$

$$\dot{p}_H = -p_H \left(2\gamma \frac{\dot{a}}{a} + (\gamma + 1)\frac{\dot{b}}{b} \right) - \frac{1}{h} \frac{V_{z0}}{a^{2\gamma} b^{\gamma+2}}. \tag{15}$$

In Equations (12)–(15), the pressures p_z, p_r, P_0, and p_H are normalized to p_{00}, which is a constant and equal to the pressure P_0 at $t = 0$. The time is normalized by the Alfvén time $t_A = l/v_{A\varphi}$, with $v_{A\varphi} = b_{\varphi 0}/(4\pi \rho_0)^{1/2}$; $\beta = c_s^2/v_{A\varphi}^2$ is the plasma beta ratio, and $c_s = (p_{00}/\rho_0)^{1/2}$ is the sound speed. Further, $h = l/H$, $S = \tau_B/\tau_A$ is the magnetic Reynolds number, and $\tau_B = 4\pi\sigma\lambda^2/c^2$ is the magnetic diffusion time.

Finally, Chargeishvili *et al.* (1993) obtain the basic equations describing the time evolution of $a(t)$, $v_\varphi(t)$, and $b(t)$ from the equations of motion. Thus, Equation (4) is written as

$$\ddot{a} = \frac{\beta}{a^{2\gamma-1} b^{\gamma-1}} - \frac{2}{ab} + a v_\varphi^2, \tag{16}$$

$$\dot{v}_\varphi = \frac{2B_1}{a^2 b} - 2\frac{\dot{a}}{a} v_\varphi, \tag{17}$$

$$\ddot{b} = \frac{\beta}{a^{2\gamma-2} b^\gamma} + G_1 b, \tag{18}$$

and for V_{z0} one obtains

$$\dot{V}_{z0} = h\beta a^2 b p_H - G_2 - V_{z0}\frac{\dot{b}}{b}, \tag{19}$$

where the time, pressure and velocity are also normalized to τ_A, p_{00} and $V_{A\varphi}$, respectively, and $G_1 = 2g_0\tau_A/R_o$, $G_2 = g_0\tau_A/V_{A\varphi}$. The equation of state for an ideal gas is written as

$$P = nkT. \tag{20}$$

This leads to the following expression for the normalized temperature

$$T = T_0 - T_1\frac{z}{\lambda} - \left(T_{1r}\frac{r^2}{\lambda^2} + T_{1z}\frac{z^2}{\lambda^2}\right), \tag{21}$$

where

$$T_0 = a^2 b P_0, \tag{22}$$

$$T_1 = a^2 b p_H h, \tag{23}$$

$$T_{1r} = \frac{1}{2a^{2\gamma} b^{\gamma-1}}, \tag{24}$$

$$T_{1z} = \frac{1}{2a^{2\gamma-2} b^{\gamma+1}}. \tag{25}$$

Using Ohm's law in the form

$$\frac{\mathbf{J}}{\sigma} = \mathbf{E} + \frac{1}{c}\mathbf{V} \times \mathbf{B}, \tag{26}$$

one obtains the normalized expressions for the components of the electric field:

$$E_r = E_{rrz}\frac{rz}{\lambda^2} + E_{rr}\frac{r}{\lambda}, \tag{27}$$

$$E_\varphi = E_{\varphi rz}\frac{rz}{\lambda^2} + E_{\varphi r}\frac{r}{\lambda}, \tag{28}$$

$$E_z = E_{zrr}\frac{r^2}{\lambda^2} + \frac{1}{Sa^2 b}, \tag{29}$$

where the electric field is normalized by $b_{\varphi 0}V_{A\varphi}/c$.

$$E_{rrz} = \frac{\dot{b}}{a^2b^2} + 2B_1\frac{v_\varphi}{a^2b}, \tag{30}$$

$$E_{rr} = \frac{V_{z0}}{a^2b} - v_\varphi B_{z0}, \tag{31}$$

$$E_{\varphi rz} = -2B_1\frac{\dot{a}}{a^3b} - B_1\frac{\dot{b}}{a^2b^2}, \tag{32}$$

$$E_{\varphi r} = B_{z0}\frac{\dot{a}}{a} - B_1\frac{V_{z0}}{a^2b}, \tag{33}$$

$$E_{zrr} = \frac{1}{a^2b}\left(B_1v_\varphi - \frac{\dot{a}}{a}\right). \tag{34}$$

Equations (8)–(25) form the full set of expressions governing the behavior of a plasma in the region of collision. It must be mentioned that Equations (16)–(18) form a complete system which is independent of the other equations but which governs their solutions. Hence, the solution of the full system of Equations (8)–(34), and consequently the time development of all physical quantities depends strongly on the parameters and the initial values of the variables which appear in the system of Equations (16)–(18). On the other hand, by changing the parameters and initial values of the variables included in Equations (8)–(34) except (16)–(18), one may expect to obtain the corresponding changes of the time histories of a limited number of physical quantities only. The most important parameter, which appears in the right hand side of Equation (16), is the value of β (see also Sakai, 1990). Its value controls the competition between the radial pressure gradient (the first term), the magnetic Lorentz force which is responsible for the pinch effect (the second term), and the centrifugal force (the third). Likewise essential is the parameter B_1 in Equation (17). It plays a role similar to β, by controlling the strength of the electric current, and it determines the ability of the magnetic Lorentz force to overcome the centrifugal force. Another important variable is the initial value of the poloidal component of the velocity. It may play an important role in the competition between the forces mentioned above. One more important parameter of the system (16)–(18) is the ratio between the initial values of the pressure gradients in the z and R directions, caused by the third term at the right hand side of the pressure expression (11). Actually, it shows how strongly a plasma can be compressed by pinching without escaping into the z-direction. The reverse effect is the excitation of a plasma jet in the z direction through escaping plasma. As to the geometry, the above parameters determine the ratio between the contact length of the two colliding loops and their diameters, in other words: the ratio between the characteristic scale-length in the radial ($L_r = |(1/p)(\partial p/\partial r)|^{-1}$)

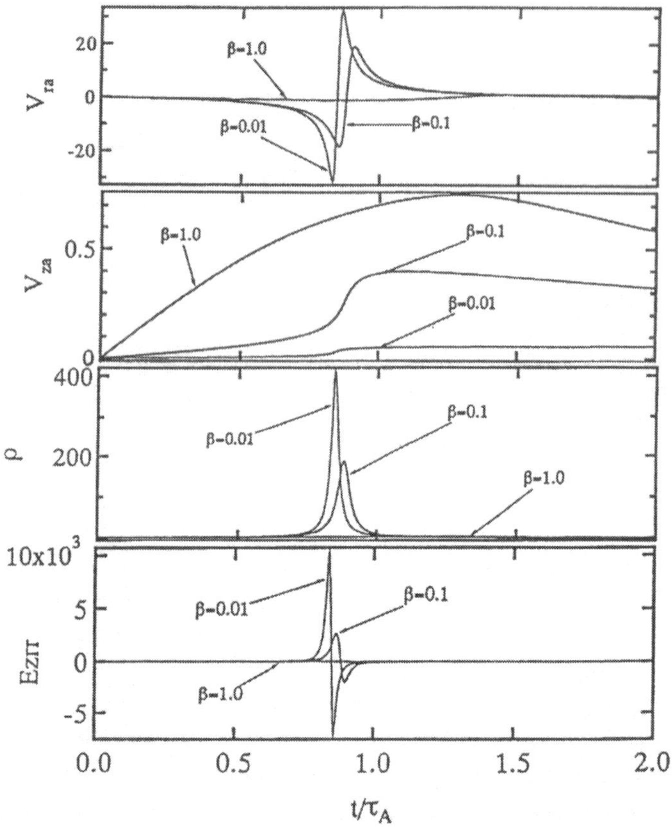

Figure 52. Y-type reconnection (L_r/L_z = 1). From *above* to *below*: time histories of the amplitudes of the normalized axial values of the radial and longitudinal components of the fluid velocity, the plasma density and the amplitude of the z-component of the electric field. The time is normalized to the Alfvén time; $V_{ra} = \dot{a}/a$ and $V_{za} = \dot{b}/b$; the electric field is normalized as stated with Equation (29). Note the strong enhancement of the physical quantities during the pinch. (Chargeishvili *et al.*, 1993. From *Solar Phys.*)

and longitudinal $(L_z = |(1/p)(\partial p/\partial z)|^{-1})$ directions. From Equation (11) we have $L_r/L_z = a_{in}^2/b_{in}^2$. On the other hand, a_{in} and b_{in} are coupled by the initial condition for the density ρ through $a_{in}^2 b_{in} = 1$, as follows from Equation (8); hence we have $L_r/L_z = b_{in}^{-3}$. The case with $L_r/L_z \gg 1$ describes a point-like collision and it is related to an X-type of two-loops interaction. However, $L_r/L_z = 1$ and $L_r/L_z \ll 1$ are typical for the Y- and I-type collisions, respectively.

We present some results (Chargeishvili *et al.*, 1993) of numerical simulations, obtained by using Equations (8)–(34). Figure 52 shows the time history of some physical quantities for three initial values of the plasma beta, for a Y-type collision $(L_r/L_z = 1)$. Clearly, the low-beta case $(\beta = 0.01)$ shows a rapid pinch with a strong initial inflow, followed by abrupt outflow of plasma. Simultaneously a strong enhancement of the amplitudes of the magnetic and electric field components is

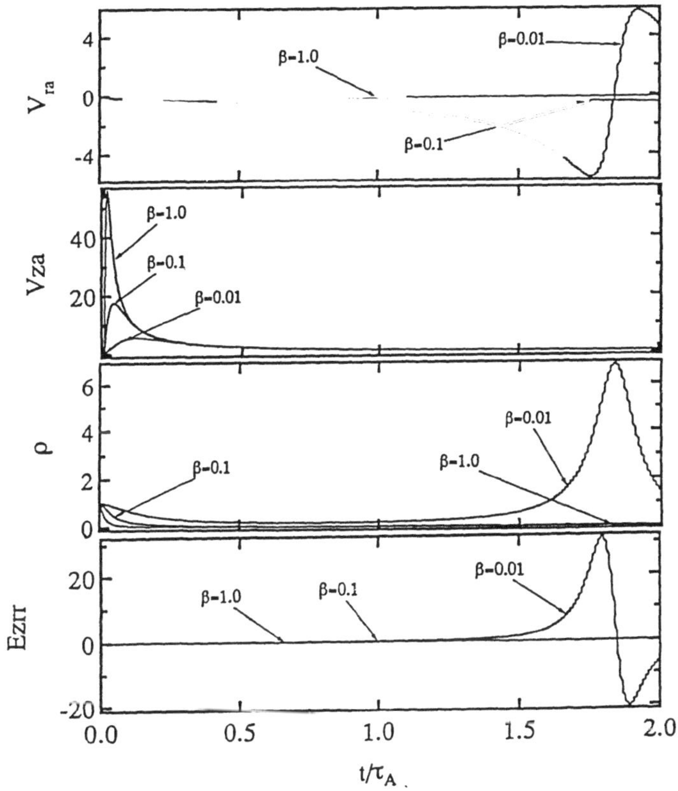

Figure 53. X-type reconnection ($L_r/L_z \gg 1$). The meaning of the symbols and the normalizations are as in Figure 52. For small beta a strong jet is produced within a time interval of 0.2 τ_A. (Chargeishvili *et al.*, 1993. From *Solar Phys.*)

observed. Likewise the temperature and density show a rapid and strong growth. An increase of β causes a weakening of the pinch effect and a decrease of the extreme values of most of the physical quantities. However, the maximum value of the amplitude of the longitudinal component of the fluid velocity increases and it becomes significant when β approaches unity.

Figure 53 refers to the case when $L_r/L_z \gg 1$, which corresponds to an X-type collision. There is apparently no strong pinch pattern but rather a strong enhancement of the longitudinal component (V_{za}) of the plasma velocity. This corresponds to strong jet production which takes place within 0.2 Alfvén transient times. Smaller β-values correspond to higher resulting values of the jet velocity.

Figure 54, to be compared with Figures 52 and 53, shows the time history of various physical quantities for the case of I-type collision, and Figure 55 combines in one graph the variation of physical quantities for the three cases of coalescence, discussed here.

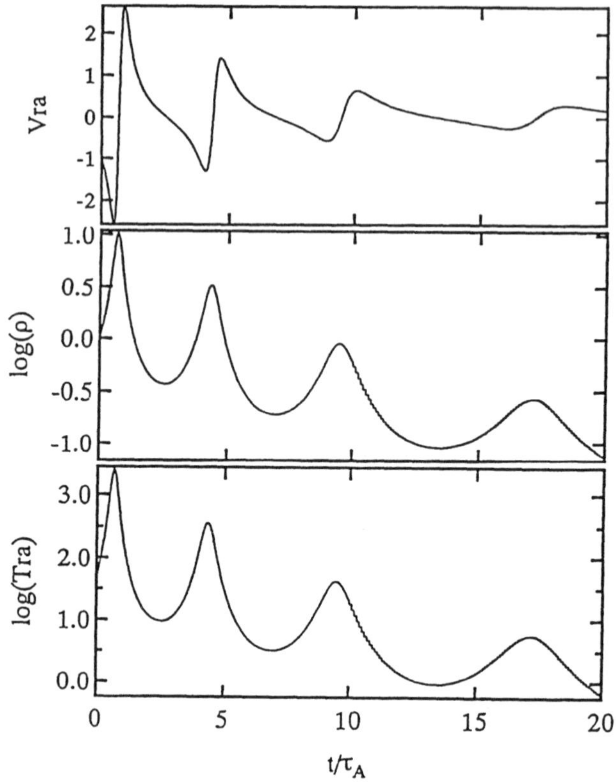

Figure 54. I-type reconnection ($L_r/L_z \ll 1$). Symbols and normalizations are as in Figure 52 and 53. Note the oscillatory character of the physical quantities. (Chargeishvili *et al.*, 1993. From *Solar Phys.*)

As demonstrated by Figure 54, the case of I-type collision ($L_r/L_z \ll 1$) reveals neither a rapid pinch nor the production of jets, and there is not even a strong enhancement of the electric field. The only interesting characteristic of this case, as shown by the long-time simulations in Figure 54, is that it has a pronounced pulsating character with a damping amplitude and a growing quasi-period with an initial value of about $3\tau_A$. Here it has to be mentioned that the case when $L_r/L_z = 1$ also shows (see Figure 56) a pulsating behavior, but with a larger decrement. When $L_r/L_z \gg 1$ there are no pulsations.

Figure 57 shows for Y-type reconnection ($L_r/L_z = 1$), a comparison between the time profiles for different initial values of the parameter B_1, a parameter that is (inversely) related to the initial current-strength. It is clear from these figures that a decrease of B_1 causes a weakening of the pinch effect and a decrease of all relevant resulting values of the physical quantities. Other simulations demonstrate that for the case $L_r/L_z \gg 1$, the value of B_1 has no influence on the production of jets.

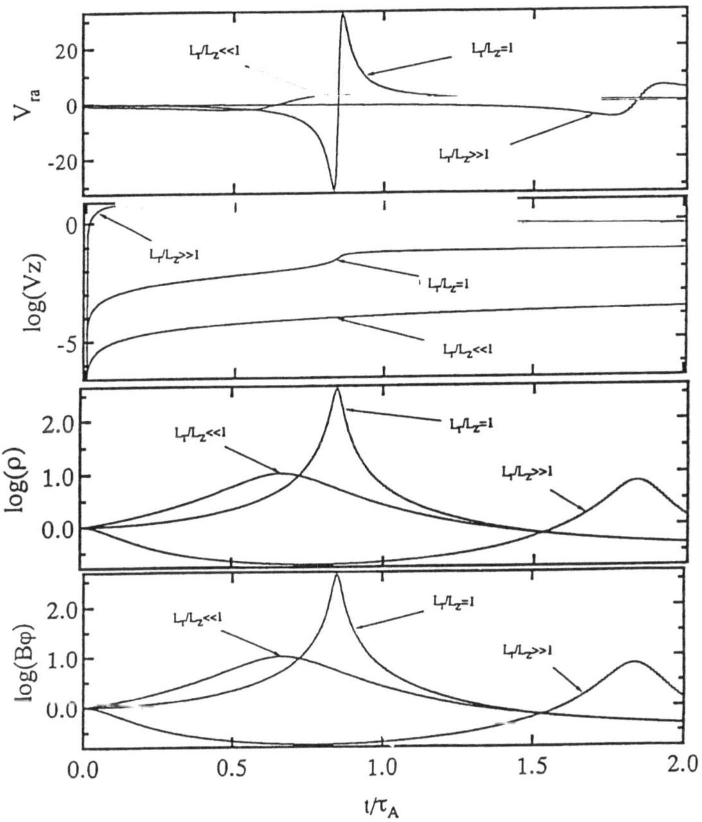

Figure 55. The influence of different initial values of the ratio L_r/L_z on the time histories of several physical quantities. The cases X, Y-, and I-type reconnection correspond with $L_r/L_z \gg$, $=$, and $\ll 1$, respectively. From *above* to *below*: the time-dependent radial and axial components of the velocity and the density and φ-component of the magnetic field. The normalizations are as in Figure 52; the magnetic field is normalized to its initial value. (Chargeishvili *et al.*, 1993. From *Solar Phys.*)

Figure 58 deals, as Figures 56, 57, and 59, also with the case $L_r/L_z = 1$, and presents a comparison between the time profiles of the radial velocity component and of the axial electric field for the cases with and without initial poloidal velocities $V_{\varphi in}$. The case $V_{\varphi in} = 0$ corresponds to a direct collision of loops and $V_{\varphi in} = 0.5$ corresponds with an oblique collision. Simulations revealed that a direct collision produces a more rapid pinch and a relatively stronger enhancement of all physical quantities. However, $V_{\varphi in}$ has no influence on the production of jets.

Figure 59 describes for $L_r/L_z = 1$ the influence of the initial condition of B_{z0} on the radial and φ-components of the electric field and the axial component of the magnetic field. The case $B_{z0in} = 0$ corresponds with complete reconnection (when both the poloidal and the longitudinal components of the magnetic field reconnect). In the case $B_{z0in} = 10$ only the poloidal component reconnects completely, while

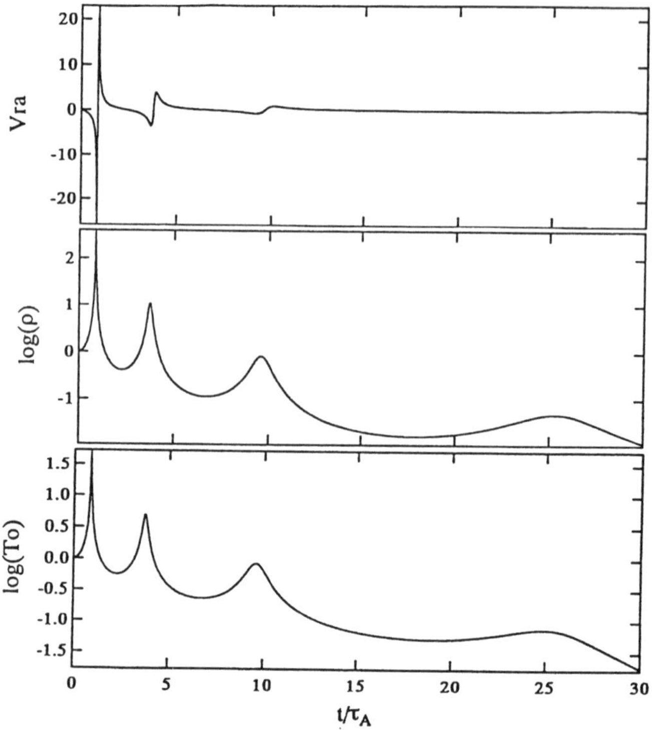

Figure 56. Oscillatory character of some physical quantities during Y-type collision ($L_r/L_z = 1$). From *above* to *below*: the radial velocity component at the axis, the density, and the axial temperature. Normalizations are as in Figure 52. (Chargeishvili *et al.*, 1993. From *Solar Phys.*)

the longitudinal component does still exist after reconnection. It refers to the case when the colliding loops have oppositely-directed but unequal longitudinal magnetic field components. During the reconnection, one of the loops, viz. the one with the strongest longitudinal magnetic field strength, absorbs the other. Simulations show that the increase of B_{z0in} does not change the time histories of most of the physical quantities, but it only causes a rather strong enhancement of the homogeneous component of the axial magnetic field b_{z0}, and of the amplitudes of the electric fields E_{rr} and $E_{\varphi r}$. This may be important for particle acceleration up to relativistic energies, even for the case $L_r/L_z \gg 1$ which, as demonstrated before, is only capable of producing jets when $B_{z0in} = 0$.

The simulations also revealed (not shown in the figures) that the parameter h, which is responsible for the gravity effect, influences but a few quantities, and then only to an insignificant degree. Neither have changes of the initial values of the radial or longitudinal velocities any significant influence on the results.

From a numerical simulation Zhao *et al.* (1993b) found that the plasma β, the type of the collision between two current-loops and the initial current-strength have significant effects on the characteristics of the resulting plasma temperature. First

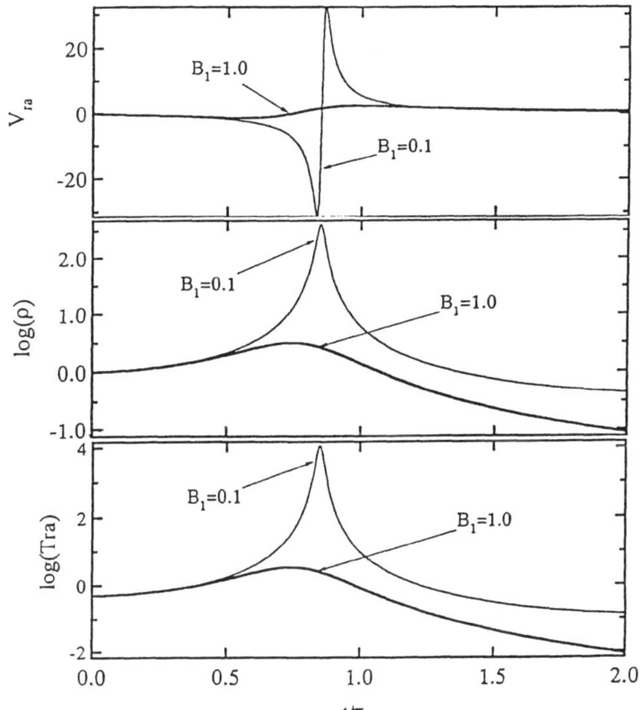

Figure 57. Y-type reconnection ($L_r/L_z = 1$). The influence of different initial values of the current-strength (determined by the parameter B_1) on the time histories of the axial values of the radial velocity component, the density, and the radial temperature component. Normalizations are as in Figure 52. (Chargeishvili *et al.*, 1993. From *Solar Phys.*)

we consider the effect of the type of collision between two current-loops on the time profile of the plasma temperature. Figures 60, 61, and 62 show the different ways the temperature profiles vary for the three types of collisions considered here. Pulsations are more apparent for I-type collisions than for the Y-type.

Next, considering the initial current strength effect, it appears that a stronger initial current leads to stronger oscillations in the time profile of the temperature (cf., Figures 60 and 61). For a weak current (case $B_1 = 0.1$) which may correspond to a weak flare, the temperature increases only by a fairly small factor, while for a strong current (case $B_1 = 10^{-4}$), which may correspond to a strong flare, the temperature can increase by factors up to a hundred. For that case, where the current has the largest relative value studied so far, the lower panel of Figure 61 shows that the time profile of the plasma temperature in an Y-type collision has five peaks, of which the third is the highest. However, there is only one single pulse in the case of an X-type collision, even though the amplitude of the peak becomes higher with increasing current-strength (as shown by the lower panel of Figure 62).

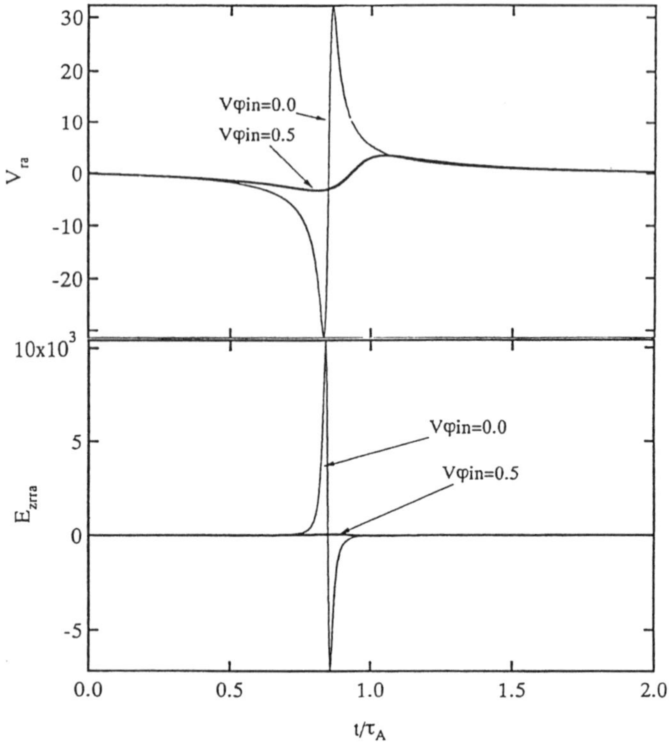

Figure 58. Y-type reconnection ($L_r/L_z = 1$). Comparison of the time histories of the normalized values of the radial velocity component and the axial electric field component for direct ($V_{\varphi in} = 0$) and oblique ($V_{\varphi in} = 0.5$) collisions. (Chargeishvili *et al.*, 1993. From *Solar Phys.*)

As to the plasma β, its increase causes an increase of the width of the first peak and of the quasi-period and a decrease of the peak amplitude. In a certain range of β values, the oscillations disappear. For weak currents this happens near $\beta = 0.02$.

The flare of 17 August, 1992, at 23:58 UT (Takahashi *et al.*, 1995) showed double peaks in the soft X-ray flux. During the first peak four brightening footpoints were observed, located along a line. They belonged to two pairs separated by the magnetic neutral line. No high-energy particles were detected; only plasma heating occurred, and a small flux of electrons with energies around 20 keV was observed before the first peak of the soft X-ray flux. Comparing the above-described numerical results with the time profile of the soft X-ray flux, we deduce that only the case of an I-type collision would be capable of producing such an event. The interval between the first and second peaks obtained from the simulation coincides well with the observed value if the flare length scale, the plasma density, and B_1 are about 10^9 cm, 10^{11} cm^{-3}, and 0.1, respectively. The maximum temperature in the first peak is about three times higher than that in the second peak. By using a particle acceleration model (cf., Zhao *et al.* 1993a), it is found that particles can

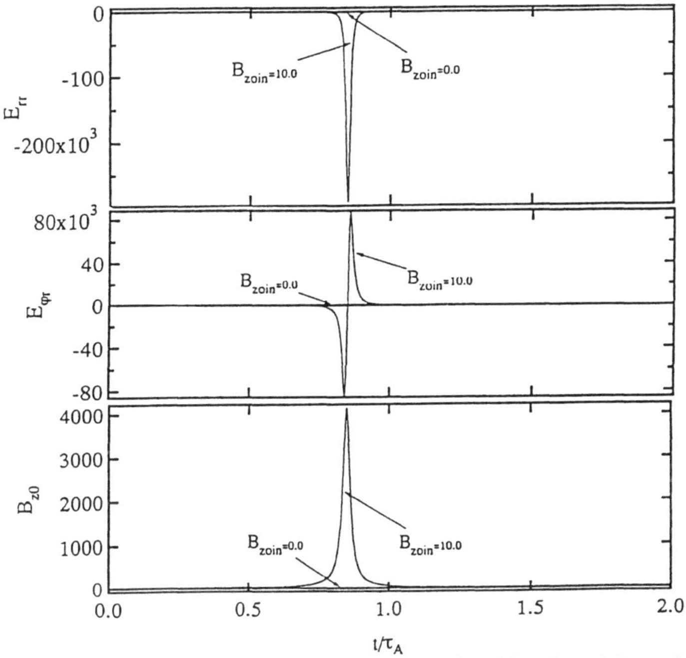

Figure 59. Y-type reconnection. Comparison between the time histories of the radial (*above*) and φ-components (*middle*) of the electric field, and the axial component of the magnetic field (*below*) for complete ($B_{z0in} = 0$) and partial ($B_{z0in} = 10$) reconnection. The normalizations are as in Figures 52 and 55. (Chargeishvili *et al.*, 1993. From *Solar Phys.*)

be accelerated up to about 30 keV within 1 s, during an interval of time preceding the first temperature maximum. There is no particle acceleration near the time of the second peak because at that time the electric field is not strong enough. For a further discussion of this flare we refer to Section 6.5.3.

4.5. High-Energy Particle Acceleration

Zhao *et al.* (1993a) studied the motion of a test proton in an electromagnetic field, by using Equations (27)–(29). The time-dependent amplitudes of the electromagnetic field are assumed to be constant during the acceleration time ($1000\ \omega_{cp}^{-1} \approx 0.001$ s, if $V_{A\varphi} = 10^8$ cm s^{-1} and $\lambda = 10^9$ cm), since that time interval is much shorter than the duration of a two-current-loops implosion ($0.01\tau_A \approx 0.1$ s if $V_{A\varphi} = 10^8$ cm s^{-1} and $\lambda = 10^9$ cm). Therefore, they investigated the motion of a test proton for the case with maximum amplitudes of the electromagnetic field components.

Figure 63 shows the time history of the Lorentz factor $\gamma = (1 + p^2)^{1/2}$ for all three cases: X-, Y-, and I-type collisions. To all three cases the same initial conditions apply: $r(0) = 1.0$, $\varphi(0) = 0.0$, $z(0) = 1.0$, $P_r = P_\varphi = P_z = 10^{-4}$, $\beta = 0.01$. As seen from this figure, proton acceleration occurs only in the case of an Y-type collision. In the cases of X- and I-type collisions, protons can not be accelerated, because the electromagnetic field components are not strong enough

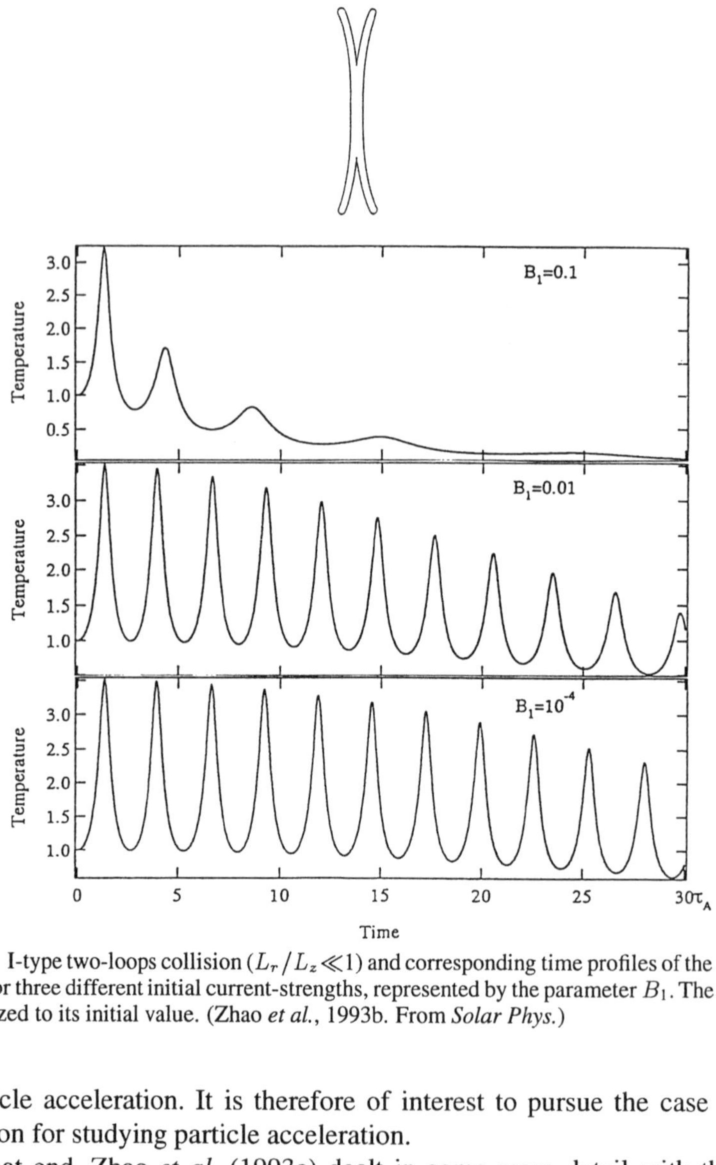

Figure 60. I-type two-loops collision ($L_r/L_z \ll 1$) and corresponding time profiles of the plasma temperature for three different initial current-strengths, represented by the parameter B_1. The temperature is normalized to its initial value. (Zhao *et al.*, 1993b. From *Solar Phys.*)

for particle acceleration. It is therefore of interest to pursue the case of Y-type interaction for studying particle acceleration.

To that end, Zhao *et al.* (1993a) dealt in some more detail with the case of Y-type collisions, using the same initial conditions as those for Figure 63, but with variable β. Figure 64 shows the time history of the Lorentz factor $\gamma = (1 + p^2)^{1/2}$ for three different values of β. It appears that the proton acceleration rate decreases for increasing β.

Next, it is useful to study the variation of the particle momentum during the process of acceleration. Figure 65 (for the same initial conditions as Figure 63) gives the time history of the momenta for the three coordinates r, φ, and z. It appears that protons are primarily accelerated in the φ and z directions. From these

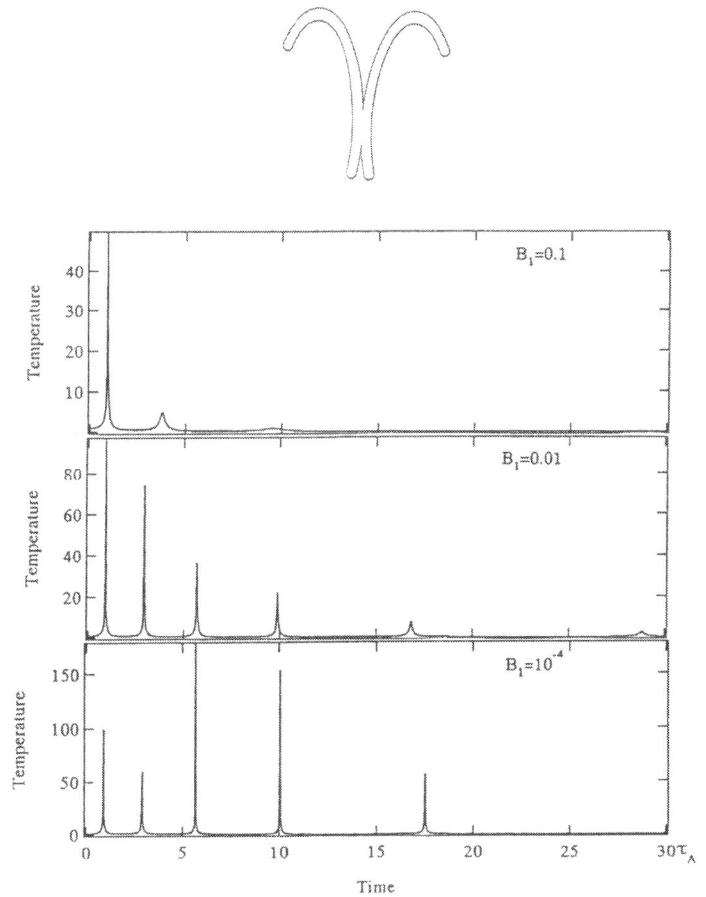

Figure 61. As Figure 60, but for Y-type two-loops collision.

and other numerical results, it is found that proton acceleration is sensitive to the initial values of the r and z coordinates as shown by Figure 66. The protons near the surface of a cylinder can be accelerated to much higher energies than those inside it. All these above numerical results are almost the same for a broad range of initial momenta, ranging from 10^{-6} to 10^{-2}.

Inserting numerical values ($V_{a,\varphi} = 10^8\,\mathrm{cm\,s^{-1}}$ and $\lambda = 10^9$), Zhao *et al.* (1993a) concluded that protons can be maximally accelerated to $\simeq 10$ GeV within a time interval of 0.001 s. For smaller initial values smaller acceleration energies are obtained. Zhao *et al.* (1993a) also obtained numerical results for electron acceleration, with initial conditions $r(0) = 1.0$, $\varphi(0) = 0.0$, $z(0) = 1.0$, $P_r = P_\varphi = P_z = 10^{-4}$, $\beta = 0.01$. Electron acceleration occurs only in the case of an Y-type collision, as shown by Figure 67. By inserting numerical data, it was concluded that a test electron can be accelerated to $\simeq$5 MeV within $1000\omega_{ce}^{-1} = 10^{-6}$ s, if

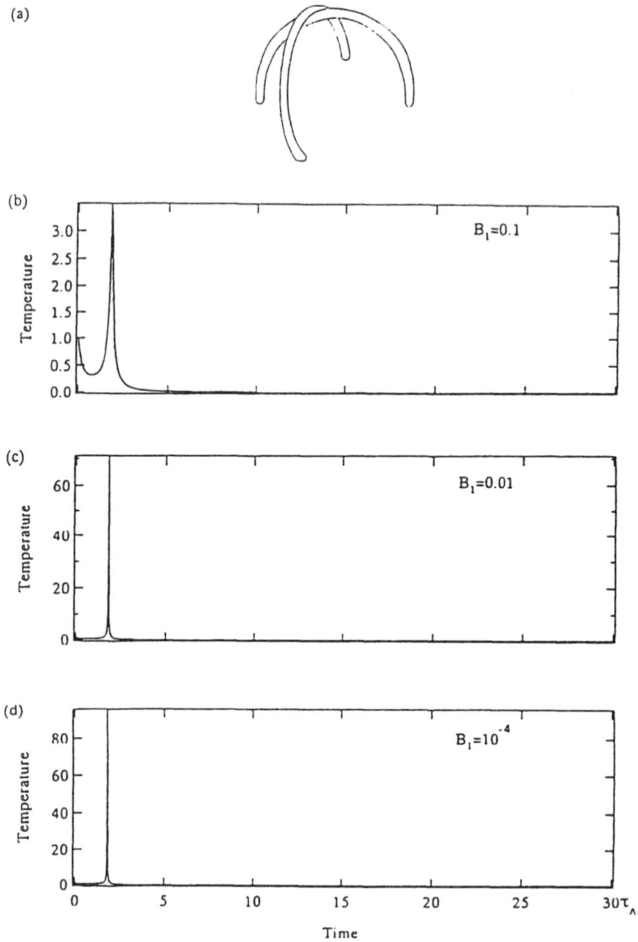

Figure 62. As Figure 60, but for X-type two-loops collision.

$V_{A\varphi} = 10^8$ cm s^{-1} and $\lambda = 10^9$ cm. Here too, smaller initial values yield smaller acceleration energies.

From the observations by the *Solar Maximum Mission* and *Hinotori* spacecraft, it became clear that there are two classes of gamma-ray/proton (GR/P) flares: the impulsive GR/P flares and the gradual GR/P flares (see, for a review, Bai and Sturrock, 1989; cf., also section 2.4 and in particular section 2.4.2 of this review). The GR/P flares are flares that are associated with nuclear gamma-rays and/or energetic interplanetary protons. Most short-lived flares correspond to the impulsive GR/P flares, while most of the long-duration flares correspond to the gradual GR/P flares. In the impulsive GR/P flares there are two phases of particle acceleration: in the first phase electrons and protons are both accelerated within 1 s to energies of $\simeq$10 MeV and $\simeq$100 MeV, respectively (Kane *et al.*, 1986; Rieger,

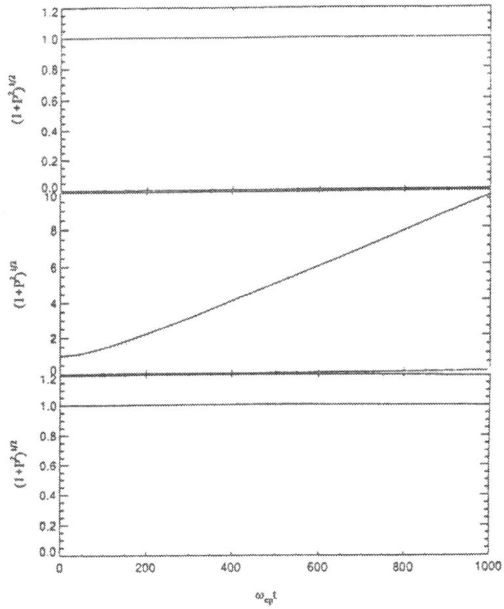

Figure 63. Time history of the proton Lorentz factor (*ordinate*) for initial conditions $r(0) = 1.0$, $\varphi(0) = 0.0$, $z(0) = 1.0$, $P_r = P_\varphi = P_z = 10^{-4}$. The time is normalized by the proton cyclotron frequency ω_{cp}. The figure shows the cases of X-type (*above*), Y-type (*middle*), and I-type reconnection (*below*). Proton acceleration appears to occur only with Y-type reconnection. (Zhao *et al.*, 1993a. From *Solar Phys.*)

1989). In the second phase electrons are accelerated up to 100 MeV and protons up to GeV energies in a time interval ranging from a few seconds to about 100 s.

In an attempt to explain the above prompt particle acceleration to relativistic energies, Sakai (1990) showed that during 3D X-type current-loop coalescence, and under suitable assumptions on the size and other physical parameters in the region of acceleration, protons and electrons may promptly (i.e., within less than 1 s) be accelerated to $\simeq$100 GeV and $\simeq$100 MeV, respectively. De Jager and Sakai (1991) showed that the duration of impulsive phase bursts (5–25 s) observed during the impulsive phases of flares can be explained quantitatively by the mechanism of 3-D X-type current-loop coalescence.

Sakai (1992) developed a model for long-duration gamma-ray/proton flares (the 'gradual GR/P flares') in order to explain prompt proton and electron acceleration during the impulsive phase. These flares are identified with the *two-ribbon flares*. Since almost all gradual GR/P flares show an impulsive behavior in the beginning of the hard X-ray emission and a gradual behavior later on (Bai and Sturrock, 1989), one identifies them with standard eruptive/dynamic flares with an impulsive phase. In most large two-ribbon flares with filament eruptions the filament begins to move up several minutes, sometimes even longer before the onset of the impulsive phase (Martin and Ramsey, 1972; Kahler *et al.*, 1988). Sakai and Koide (1992) presented a theory of filament eruption taking place before the impulsive phase of solar flares.

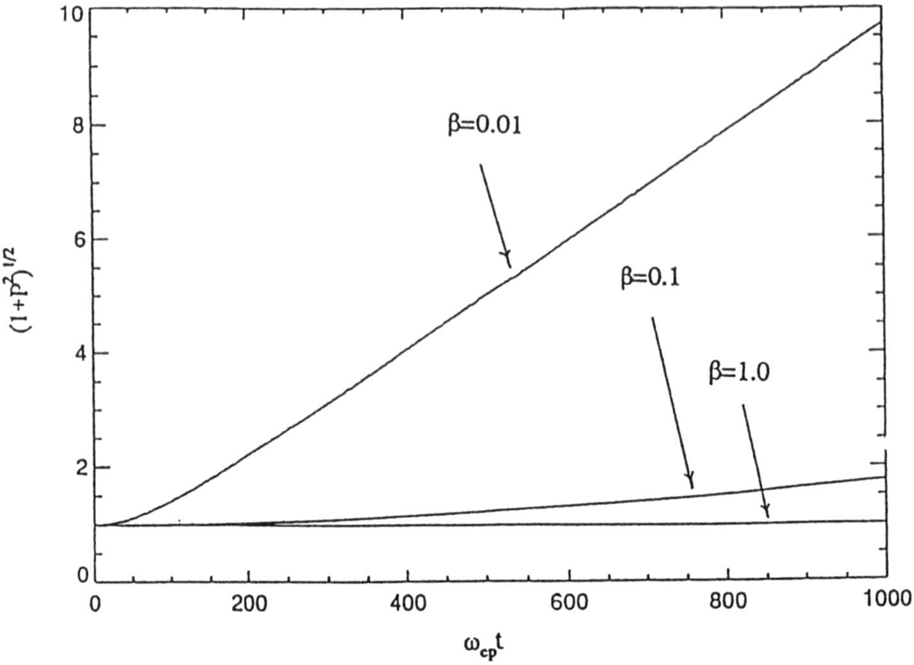

Figure 64. Comparison of the time histories of the proton Lorentz factor (*ordinate*) for different β-values in the case of Y-type collisions with the same initial conditions as for Figure 63. (Zhao *et al.*, 1993a. From *Solar Phys.*)

They could indeed demonstrate in their theoretical model that the upward motion of the X-point, which may trace the filament eruption, begins several minutes before the impulsive phase, when the magnetic field produced by the current near the X-point exceeds the magnetic field along the filament. Sakai (1992) determined the electromagnetic fields during the implosion phase of the current-sheet from the MHD equations, and investigated the motion of test protons and electrons. He found that, under reasonable asumptions about the size and velocities in the reconnection area, protons and electrons can both be accelerated promptly: within 1 s to $\simeq$70 MeV and $\simeq$200 MeV, respectively.

4.6. Shock-Wave Formation during Current-Loop Coalescence

In this section we present results on the generation of fast magnetosonic waves for various cases: from an unstable current-sheet, and from a colliding set of two current-loops. We also discuss fast magnetosonic shock formation during coalescence of two current-loops.

4.6.1. *Fast Magnetosonic Waves Emitted from an Unstable Current-Sheet*

The modulational instability associated with finite amplitude waves in plasmas may play an important role in various kinds of nonlinear phenomena such as

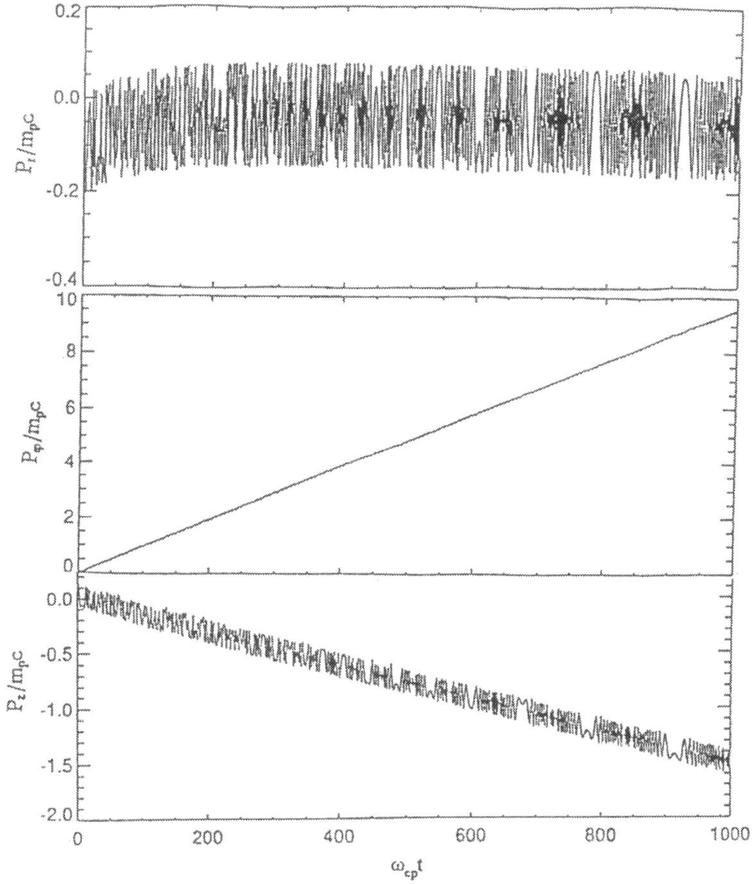

Figure 65. Time history of the normalized values of the proton momentum in three directions, for an interaction situation under the same initial conditions as for Figure 63. The figure presents data for P_r *(above)*, P_φ *(middle)*, and P_z *(below)*. (Zhao *et al.*, 1993a. From *Solar Phys.*)

the occurrence of solitons and parametric interactions among the waves. Sakai (1983) showed that transverse amplitude modulations of fast magnetosonic waves propagating perpendicularly to the ambient magnetic field are unstable and can generate slow waves, which cause plasma heating by Landau damping. He applied his results to explain the slow-wave motions observed in X-ray images taken from *Skylab* (Rust and Švestka, 1979). The observed velocities of these waves decreased in $\simeq$ 2 hours from 500 to 50 km s^{-1}. The emission process of the fast magnetosonic waves in the solar plasma is important in relation to coronal heating and shock-wave generation from a solar flare region. Koide and Sakai (1994) demonstrated, by using a MHD code, that during collisions between two current-loops fast magnetosonic shock-waves as well as finite amplitude fast magnetosonic waves can be produced.

In soft X-ray images taken by *Yohkoh*, Shimizu *et al.* (1992) found many events showing coronal transient brightening in active regions, which originated from single loops as well as from two-loops collisions. The coronal loop heating

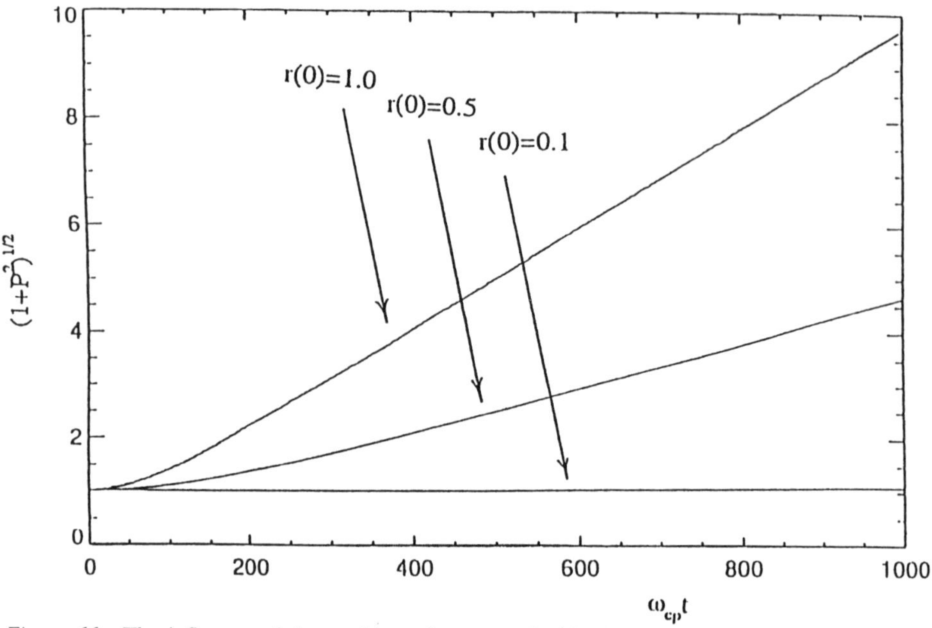

Figure 66. The influence of the position of a proton inside the reconnection area on the time-development of the proton momentum, represented here by the dependence of the normalized value of the proton momentum on normalized time. Protons at the surface of the loop appear to be accelerated strongest. The calculations were made for the same initial conditions as those for Figure 63. (Zhao *et al.*, 1993a. From *Solar Phys.*)

in active regions seems to be a rapid process, sometimes occurring in less than one minute. This observation of rapid loop heating stimulated a reconsideration of the problem of the coronal loop heating mechanism. The current-sheets and current-loops that may be produced by shearing motions at the photospheric level are not in equilibrium, especially not in the early stage of the production of the current-system. Therefore it is useful to study the dynamical processes in which a new state of equilibrium is approached, starting from a non-equilibrium state with current-sheets or current-loops. Following such considerations, Sakai *et al.* (1994) and Nishikawa *et al.* (1994) investigated the loop heating process as well as the two-loops coalescence process by using a 3-D electromagnetic particle code.

Fushiki and Sakai (1994) investigated how fast magnetosonic waves can be produced from a pinching current-sheet, by using a 3-D MHD code. They found that, following the pinching of the current-sheet due to a pressure unbalance, the current-sheet begins to expand due to the excess of plasma pressure in the center of the current-sheet. During the expansion phase, strong fast magnetosonic waves can be created in the region of a steep density gradient and they propagate away from the current-sheet. The fast magnetosonic waves become unstable against the modulational instability as found by Sakai (1983). After the emission of the fast

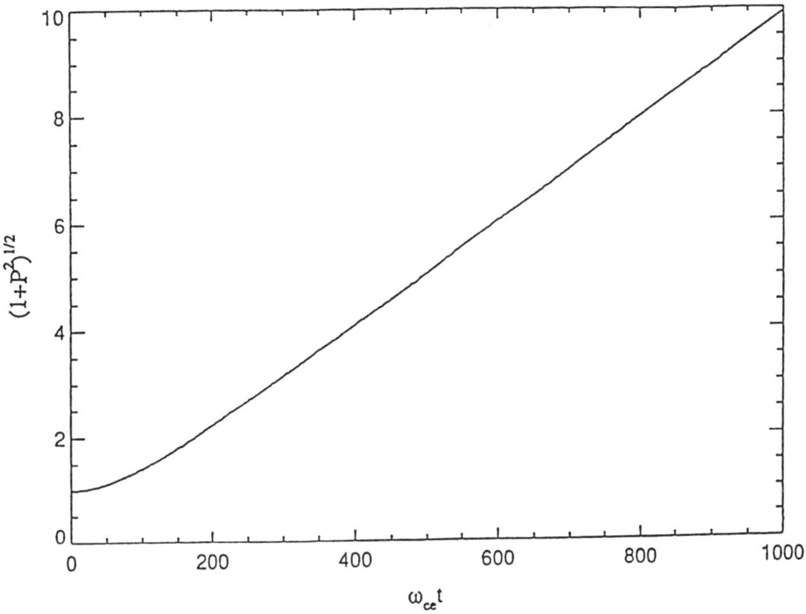

Figure 67. Time history of the electron Lorentz factor (*ordinate*) for a Y-type collision under initial conditions: $r(0) = 1.0$, $\varphi(0) = 0.0$, $z(0) = 1.0$, and $P_r = P_\varphi = P_z = 10^{-4}$. Electron acceleration, like that of protons, appears to be efficient only in the case of Y-type interaction. (Zhao *et al.*, 1993a. From *Solar Phys.*)

magnetosonic waves the current-sheet will relax to a new equilibrium state, where the current-sheet can be heated by adiabatic compression.

Fushiki and Sakai (1994) used a 3-dimensional simulation code of the MHD equations, in which the numerical scheme is a modified Lax–Wendroff method. The system sizes, L_x, L_y and L_z in the x, y, and z directions are respectively $0 \leq L_x = L_y \leq 2\pi$ and $-\pi \leq L_z \leq \pi$. The number of mesh-points, N_x, N_y, N_z in the x, y, and z directions are $N_x = 5$, $N_y = 32$, $N_z = 201$. The time step is 0.003 Alfvén times. The boundary conditions are free for the x- and z-axes and are periodic along the y-axis. The velocity is normalized by the Alfvén velocity, $v_A = B_0/(4\pi\rho)^{1/2}$. The density, the magnetic field and the pressure are normalized by ρ_0, B_0 and ρ_0. The time is normalized by $\tau_A = L_z/v_A$. The space is normalized by L_z. An important parameter is the plasma beta $\beta = c_s^2/v_A^2$, where $c_s^2 = p_0/\rho_0$.

The plasma configuration with a magnetic neutral sheet is important for the magnetic energy release processes related with coronal transient brightenings and solar flares. The interaction between a pre-existing magnetic field and the emerging flux may produce a current-sheet, where the sheet is not in a state of equilibrium in the early stages of formation of the current-sheet. A current-sheet in the low-beta plasma of the solar corona will pinch, due to the magnetic force. For this configuration we study how a non-equilibrium current-sheet will relax to a new

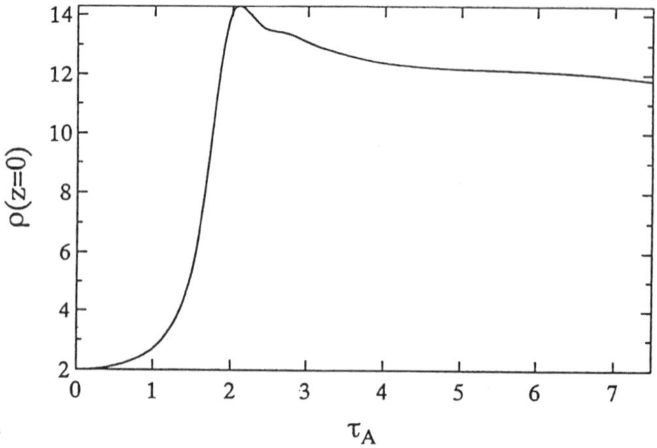

Figure 68. Current-sheet formation. Time history of the density in the center of a non-equilibrium current-sheet ($x = y = z = 0$) with an imposed small density-perturbation. (Fushiki and Sakai, 1994. From *Solar Phys.*)

state of equilibrium. For a non-equilibrium current-sheet the relative values of the density, pressure, and magnetic field are taken as follows:

$$\rho = p = 1 + \frac{1}{\cosh^2(z)}, \tag{35}$$

$$B_y = \tanh(z). \tag{36}$$

For the plasma beta the value $\beta = 0.01$ is chosen. A very small density perturbation is imposed, chosen as

$$\Delta\rho = 10^{-5}\sin(y). \tag{37}$$

The reason why the above density perturbation is imposed, as a seed perturbation, is to investigate the modulational instability of fast magnetosonic waves propagating in the z-direction. The simulation results are obtained for the initial conditions of Equations (35)–(37). Figure 68 shows the time development of the density at the center of the current-sheet. The density appears to increase by about a factor of seven around $t = 2\tau_A$. Thereafter, the current-sheet expands due to the enhancement of the plasma pressure.

Figure 69 shows the time-history and spatial distribution in the z-direction (at $x = y = 0$) of the density and the magnetic field, B_y, at the time points $t = 0$, 1.872, and $2.34\tau_A$. Apparently, during the pinching phase the density and magnetic field gradients become steeper. The plasma expansion starts from the central region of the sheet.

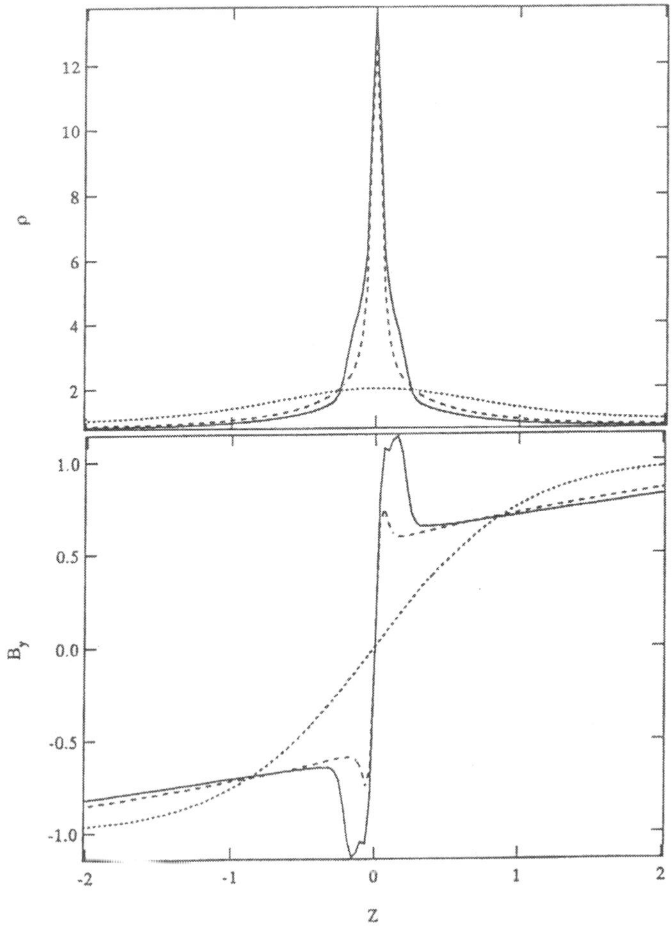

Figure 69. Current-sheet formation; the case of an imposed small density-perturbation (cf., Figure 68). Density (*above*) and magnetic field B_φ (*below*) in the z-direction taken at three different times: t/τ_A = 0 (dotted), 1.872 (dashed) and 2.34 (solid). Density and field are normalized to their initial values. (Fushiki and Sakai, 1994. From *Solar Phys.*)

As seen in Figure 70, fast magnetosonic waves can originate in the region of plasma expansion, and they propagate away from the current-sheet. Figure 70 shows the time history and the spatial distribution in the z-direction (for $x = y = 0$) of the density ρ and magnetic field, B_y, at time points $t = 2.808, 4.21$, and $5.616\tau_A$.

It appears from the B_y figure for $t = 5.616\tau_A$ (right-hand lower part of Figure 70), that a new current-sheet can be formed after the emission of the fast magnetosonic waves. Figure 71 gives a three-dimensional presentation of the density and the magnetic field B_y at $t = 4.21\tau_A$.

To study the characteristics of the waves emitted from a current-sheet, Fushiki and Sakai (1994) analyzed the density distribution during the period $3.021\tau_A$ to $6.64\tau_A$ in the region $z = 0.942$–2.92 (for $x = y = 0$), where 64 data points are

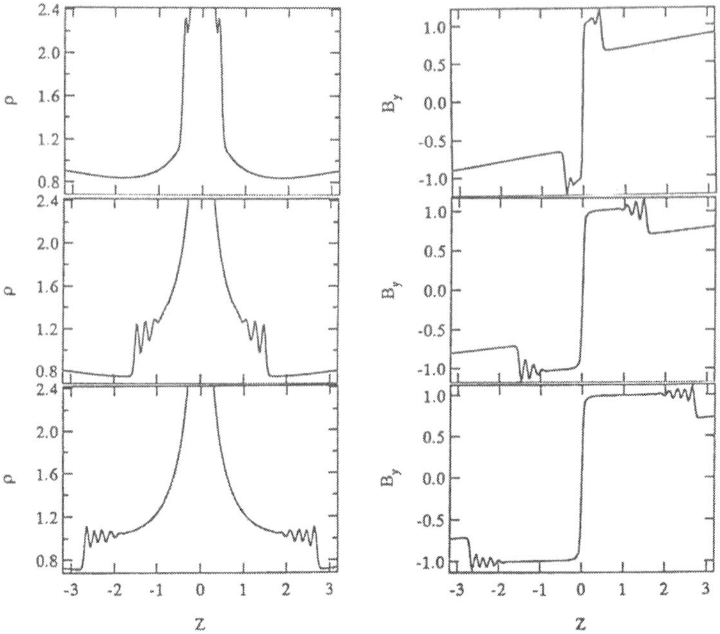

Figure 70. Wave emission from a current-sheet. Time history of density (*left*) and magnetic field, B_y (*right*) in the z-direction (at $x = y = 0$), taken at three different times, $t/\tau_A = 2.808$ (*above*), 4.21 (*middle*), and 5.616 (*below*). Normalizations are as in Figure 69. (Fushiki and Sakai, 1994. From *Solar Phys.*)

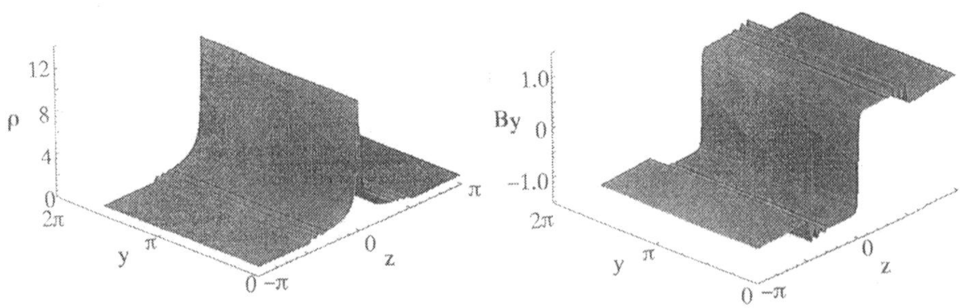

Figure 71. Wave emission from a current-sheet. Three-dimensional plots of the density (*left*) and magnetic field B_y (*right*) at $t = 4.2\tau_A$. Normalizations are as in Figure 69. (Fushiki and Sakai, 1994. From *Solar Phys.*)

taken. From the time- and z-dependence of the density they performed a two-dimensional Fourier transformation (one coordinate in the z-direction and one in time), and thus they obtained the dispersion relation of the density fluctuations, shown in Figure 72. In this figure, the frequency ω is normalized by τ_A, and the wave number k_z is normalized by L_z. The strongest emission appears in the low-frequency part, which may correspond to the slowly varying background density. A second strong emission part is found with its peak intensity at $\omega = 24$ and $k_z = 30$. The strong emission part shows that magnetosonic waves were created

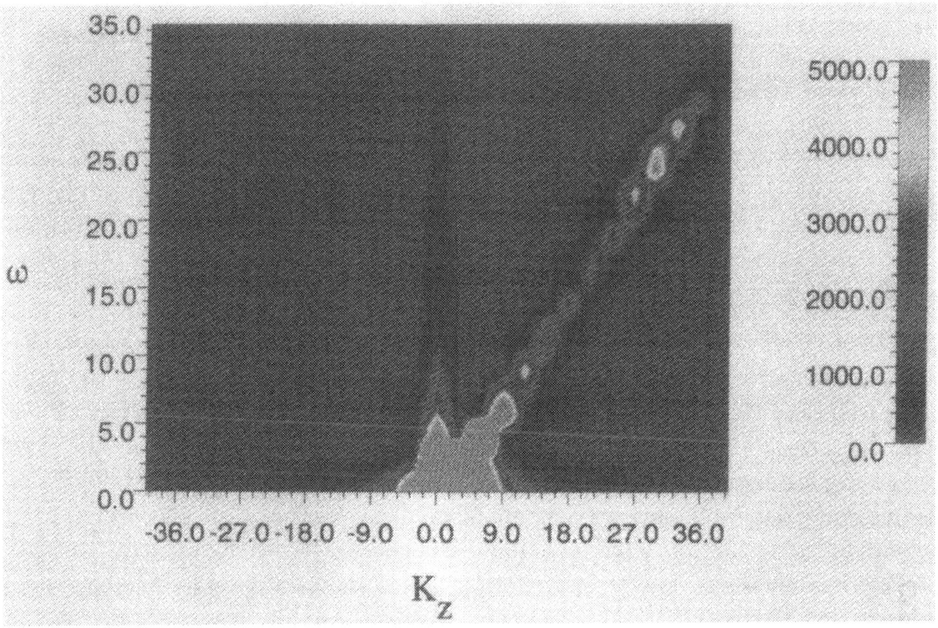

Figure 72. Wave emission from a current-sheet. Dispersion relation of the density perturbation in the z-direction. The diagram is produced from data on the density-time variation during the period (3.021–$6.64\tau_A$) and applies to the z-interval (0.942–2.92). The frequency ω is normalized by τ_A, and k_z is normalized by L_z. (Fushiki and Sakai, 1994. From *Solar Phys.*)

in the current-sheet, and propagate in the positive z-direction. The wavelength calculated from $k_z = 30$ is approximately equal to the wavelength of the density perturbations observed in Figure 70. The propagation velocity of the waves is $\omega/k_z = 0.8v_A$. Therefore, the waves emitted from the current-sheet must be identified as fast magnetosonic waves. These fast magnetosonic waves decay during their propagation out of the emission region. The most important decay mechanism for the fast magnetosonic waves is a modulational instability, as predicted by Sakai (1983). If the wave-amplitude of the fast magnetosonic waves exceeds a critical value, given by $\rho/\rho_0 = \sqrt{\beta}$, the fast magnetosonic waves become unstable against transverse modulation along the ambient magnetic field and they generate slow waves propagating along the magnetic field.

In order to study the properties of the slow waves, Fushiki and Sakai (1994) calculated the density perturbation, $\Delta\rho = \rho(y, z, t) - \rho(y = 0, z, t)$ at $x = 0$. Figure 73 shows the time history of the density perturbation associated with the slow waves. The seed density perturbation appears to increase by about a factor ten, due to the modulational instability. The theoretical growth-rate (Sakai, 1983) is given by $\gamma\tau_A = k_y v_A \tau_A \phi$ where ϕ is the normalized fast wave amplitude and k_y the wave number in the direction along the magnetic field. If we take for the observed wave amplitude ϕ the value $\phi = 0.7$, and use the dispersion relation

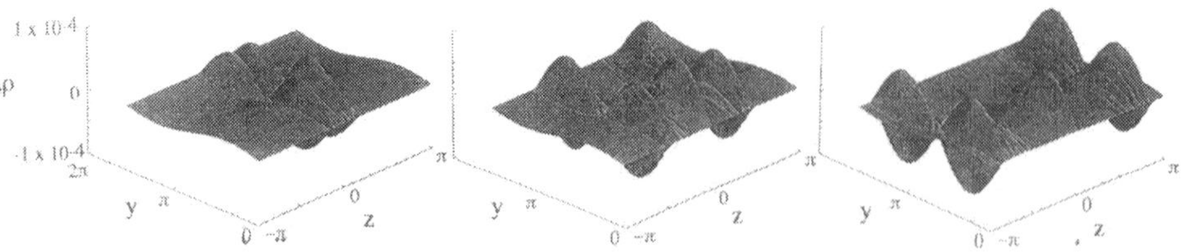

Figure 73. Slow-wave excitation. Three-dimensional plots of the density perturbation taken at $x = 0$, while $\Delta\rho = \rho(y, z, t) - \rho(y = 0, z, t)$ is the perturbation associated with the slow waves at three different times: $t/\tau_A = 2.808$ (*left*), 4.21 (*middle*), and 5.616 (*right*). (Fushiki and Sakai, 1994. From *Solar Phys.*)

for the fast waves, one finds a growth rate $\gamma\tau_A = 0.8$, a value that is too high for explaining the observed enhancement of the slow density perturbations. If we take the interaction between the slow waves and particles into account, the slow waves can be damped by Landau damping. Therefore the fast magnetosonic waves propagating perpendicular to the magnetic field can produce slow waves due to the modulational instability, which eventually can heat the plasma by Landau damping.

The current-sheet, newly formed after the emission of the fast magnetosonic waves, can be heated by adiabatic compression. For the present situation the density at the center of the sheet can be increased by about a factor six, while the temperature is increased by about a factor 3. Such a temperature enhancement can be observed in soft X-ray images taken with *Yohkoh*. The time scale of the temperature enhancement is a few Alfvén transit times, which is less than about one minute in typical coronal conditions in solar active regions.

The processes by which fast magnetosonic waves are emitted from a pinching current-sheet, and the modulational instability of the fast waves are important for understanding matters like coronal transient brightenings, rapid loop heating, and shock-wave emission from the region of magnetic energy release where the emerging magnetic flux interacts with the pre-existing magnetic field.

Fast magnetosonic waves may be trapped in regions of enhanced plasma density. Such trapped pulsations (Nakariakov and Roberts (1995) have been invoked as explanations for a variety of periodic and quasi-periodic events observed in the corona. We refer to, e.g., Aschwanden (1987), Edwin (1991), and Bray *et al.* (1991).

4.6.2. *Shock Formation During Two-Loops Coalescence*

One of the important unsolved problems of solar flare physics is that of the mechanisms for accelerating high-energy particles during and after flares. We refer to de Jager (1988) and Sakai and de Jager (1991) who proposed that the high-energy particles observed in interplanetary space after flares can be accelerated in three phases that correspond to three different acceleration mechanisms. In the first phase, during rapid coalescence of two current-loops, it is expected that protons and electrons can both be accelerated in d.c. electric fields up to $\simeq$70 MeV and $\simeq$200 MeV, respectively, within less than 1 s (Sakai, 1992). In the second phase, fast magnetosonic shock-waves, which are thought to be produced in the first phase, can rapidly accelerate both electrons and protons to relativistic energies by the strong coherent electric fields in the shock-front (Sakai and Ohsawa, 1987; Ohsawa and Sakai, 1987, 1988a, b). This second phase follows the first after a brief delay, which may be identified as the time needed for the formation of the fast magnetosonic shocks. In the third phase, high-energy solar cosmic rays are produced more gradually by statistical acceleration (Ramaty and Murphy, 1987) in an extended shock-front system which interacts with turbulent waves, higher up in the corona.

The generation mechanism of the fast magnetosonic shock-waves in the time interval between the first and second phases of acceleration has not yet been studied in detail. Koide and Sakai (1994) presented results on shock formation during the collision of two current-loops, using a numerical simulation of a magneto-hydrodynamical (MHD) model. A three-dimensional time-dependent nonlinear compressible resistive MHD simulation code in cylindrical geometry was used (see for details Koide, 1990).

It appears that fast magnetosonic shock-waves can be generated, that run into a direction perpendicular to the line along which the two loops collide, but to that end the current in at least one of the loops should be strong enough. The time needed for shock formation is almost the same as the period in which fast magnetosonic waves propagate from one current-loop to the other.

The magnetic Reynolds number and the total parallel current density are important parameters for controlling the process of shock-wave formation during the collision of current-loops. For small magnetic Reynolds numbers ($S = 3\times10^3$) the current-loops coalesce monotonously into one current-loop without shock formation. If the current is not strong enough (e.g., $J_{z0} = 5.0$), magnetosonic waves that do not develop into shocks are generated during the collision of the two current-loops. Detailed calculations were made for the case in which the magnetic Reynolds number is $S = 3\times10^4$ and the current-density is so high ($J_{z0} = 15.0$) that the external magnetic field and the magnetic field induced by the currents are comparable. In that case a rarefaction region was found to occur at the upper and lower sides of the high-density region formed in the early stage. From there, rarefaction waves propagate into a direction opposite to that of the displacements of the current-loops, with a group velocity of $0.95 \pm 0.03 v_{\mathrm{A}}$. The waves are damped quickly. After the generation of the rarefaction wave, fast magnetosonic waves are produced and propagate from $t = 0.5\tau_{\mathrm{A}}$ into four directions. This time corresponds to the time needed for a magnetosonic wave to travel the distance between the initial current-loop position and the center of the interacting region. The propagation velocity v_p of the waves into the lateral direction is equal to 1.06 ± 0.05 (in terms of v_{A}), which generally exceeds the fast magnetosonic wave-speed $v_f = (c_s^2 + v_{\mathrm{A}}^2)^{1/2} = v_{\mathrm{A}}(1+\beta)^{1/2}$. Hence the waves become shocks. Here, c_s, v_{A} and β are the sound speed, the Alfvén velocity and c_s^2/v_{A}^2. When $J_{z0} = 5$, a similar structure is observed without shock-waves. The shock-wave catches up the rarefaction wave.

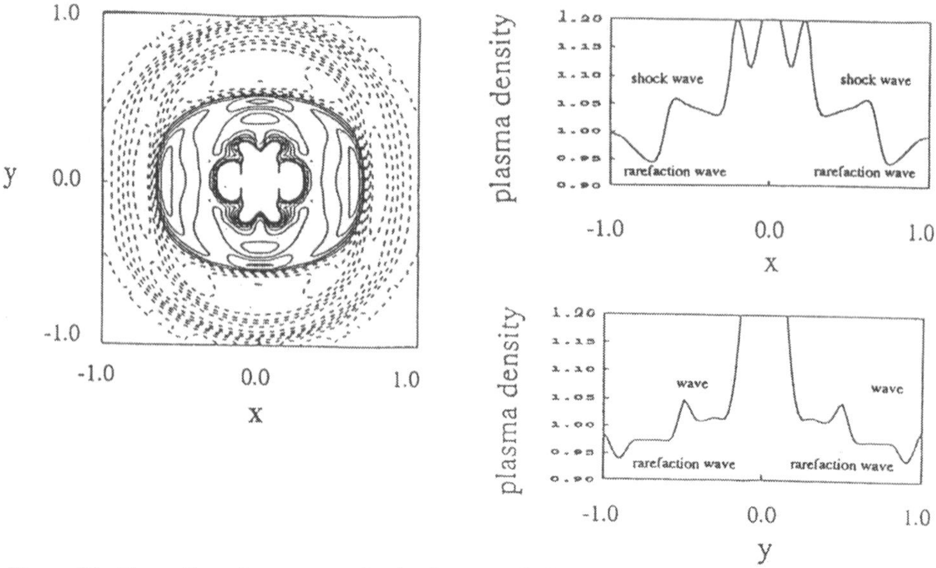

Figure 74. Generation of magnetosonic shock-waves during a collision of two current-loops. The Figure gives values of the plasma densities at the time $t = 0.7\tau_A$. *Left*: a contour map of the density (normalized to its initial value) across the current-loop. Thick lines denote unity value. *Right, above*: densities at $y = 0.0$. *Right, below*: density at $x = 0.0$. (Koide and Sakai, 1994. From *Solar Phys.*)

As can be derived from Figure 74, the ratio of the peak density and the density existing before the origin of the rarefaction wave is 1.059±0.002. This value is in agreement with the Rankine–Hugoniot conditions, which state that the density-ratio equals the ratio between the speed of propagation of the wave and the fast magnetosonic wave speed, which is $v_p/v_f = 1.055\pm0.05$. There are also fast magnetosonic waves running upward and downward with a group velocity of 0.88±0.08. In these directions the fast magnetosonic wave and the rarefaction wave are clearly separated, as appears from Figure 74 (right, below).

4.7. Collision between a Current-Loop and a Plasmoid

Another important sofar unsolved problem in solar flare mechanisms is that of flare triggering. Simnett (1993) reviewed the various coronal, chromospheric and photospheric events that produce flares, in which he included a discussion of the radio, Hα, UV and X-ray regions of the spectrum, along with magnetographic data.

A few triggering mechanisms were presented: (1) triggering by fast magnetosonic waves or shocks (Vorpahl, 1976; Sakai and Washimi, 1982; Sakai, 1982, 1983; Sakai and Nishikawa, 1983; Sakai *et al.*, 1984; Simnett *et al.*, 1990); (2) triggering by high-energy electron beams (Nakajima *et al.*, 1985); (3) triggering by two-loops collisions (Gold and Hoyle, 1960), and the interaction of emerging flux with an overlying flux-tube (Heyvaerts *et al.*, 1977; Forbes and Priest, 1984; Zirin and Tang, 1991); (4) triggering by the upward motion of a filament (Mikic *et al.*,

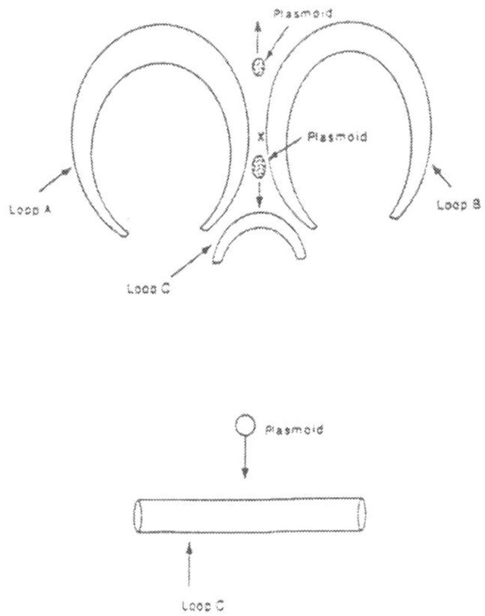

Figure 75. Above: schematic picture of the triggering of a solar flare, including plasmoid formation. *Below*: theoretical model. (Sakai *et al.*, 1995. From *Solar Phys.*)

1988; Sakai and Koide, 1992); (5) interaction between force-free current-loops and plasmoids that can emerge from the current-loop coalescence region, especially in association with the triggering of another flare (Sakai *et al.*, 1995).

Figure 75 presents a schematic picture of the above fifth triggering model of a solar flare. Let us assume, as in the model of Figure 17, that a primary flare is triggered by loop-coalescence (loops A and B). This flare produces two plasmoids that are emitted from the region of coalescence. The plasmoids move away from the primary flare region and may collide with another current-loop (C) or with several loops.

Sakai *et al.* (1995) investigated a collision process between a force-free current-loop and a plasmoid, by using a 3-D resistive MHD code (lower part of Figure 75). They found that a current-system can be induced in the front of the plasmoid when it approaches a force-free current-loop. This secondary current produced in the front of the plasmoid separates from the plasmoid and coalesces with the force-free current-loop associated with the process of magnetic reconnection. The core of the plasmoid stays outside the region of reconnection, and maintains a high density. The core can be confined by the current-system produced around the plasmoid. This collision process between a current-loop and a plasmoid may explain the triggering of certain solar flares observed by *Yohkoh* (see section 6.7.3).

Figures 76 and 77 show the time development of the isosurfaces with a current-density of $j_y = 0.003$. The plasmoid is not yet surrounded by a current-system in the early phase. However, as it approaches the current-loop, a current-system can be induced near the front of the plasmoid, as seen in the left-hand panel of Figure 76. Thereafter, the current induced at the front of the plasmoid collides with

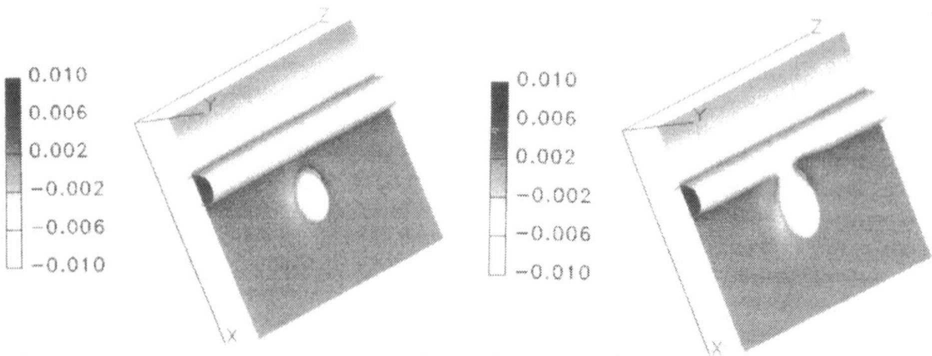

faces of current-intensity $j_y = 0.003$ and a contour of j_y on the plane $(y = \pi L_0)$ at $t = 2\tau_A$ (*left*) and $t = 3\tau_A$ (*right*). (Sakai *et al.*, 1995. From *Solar Phys.*)

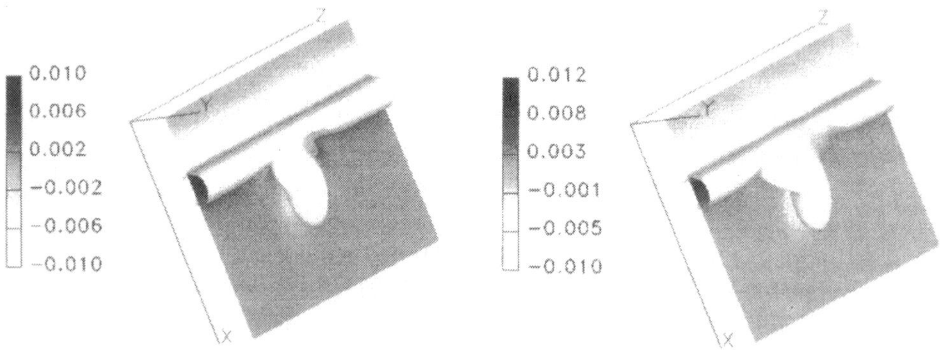

Figure 77. As Figure 76, but for $t = 4\tau_A$ (*left*) and $t = 5\tau_A$ (*right*).

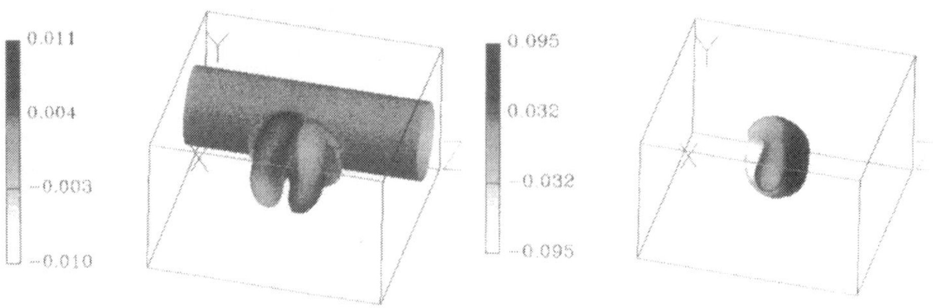

Figure 78. Collision between a current-loop and a plasmoid; cf., Figures 76 and 77. *Left*: isosurface of the total current-intensity with $|\mathbf{J}| = 0.00326$ and current density j_y over the surface at $t = 4\tau_A$. *Right*: isosurface of the total velocity for $|\mathbf{V}| = 0.106$ and the velocity distribution V_z over the surface at $t = 4\tau_A$. (Sakai, Fushiki, and Nishikawa, 1995. From *Solar Phys.*)

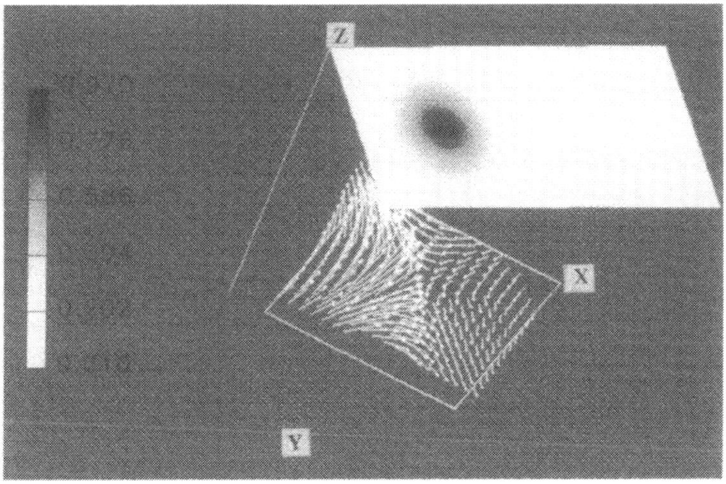

Figure 79. Collision between a current-loop and a plasmoid; cf., Figures 76–78. Vector plots of magnetic fields near the coalescence region between a force-free current-loop and a secondary current induced by the plasmoid, and a contour map of B_z on the plane $z = 4\pi L_0$ at $t = 4\tau_A$. The gray-scale panel gives the strength of the z-component of the magnetic field. (Sakai *et al.*, 1995. From *Solar Phys.*)

the force-free current-loop and deforms it, as was already seen in Figure 76 and better in Figure 77. The left-hand panel of Figure 78 shows an isosurface of the total current-intensity with $|\mathbf{J}| = 0.00326$. The gray-scale map on the surface shows the current-distribution, j_y. The Figure shows that a current-system originates with a 3-dimensional structure around the plasmoid. The right-hand panel in Figure 78 shows an isosurface of the total velocity-intensity with $|\mathbf{V}| = 0.106$, and the gray-scale map on the surface gives the velocity distribution V_z. As it appears, the core-part of the plasmoid does not move, while near the region of the collision plasma moves away from both sides of the region of interaction.

Figure 79 shows vector plots of the magnetic fields near the region of interaction and the gray-scale map shows the magnetic field distribution of B_z on the plane $z = 4\pi L_0$. It appears that an X-type magnetic neutral point originates in the interaction region between the force-free current-loop and the current induced from the plasmoid.

From the above results we conclude that coalescence between a force-free current-loop and the current induced in front of a plasmoid can lead to magnetic reconnection. This reconnection plays an important part in the process of magnetic energy release and may eventually trigger a solar flare.

5. Current-Loops Coalescence – Kinetic Effects

5.1. Single-Loop Heating – Buneman Instability and Kink Instability

One of the major problems in solar plasma physics is how the solar corona can be heated up to and above 10^6 K. It is now generally accepted that coronal heating is strongly related to the coupling of magnetic structures with the convective motions in and beneath the photosphere of the Sun. The solar photospheric and chromospheric regions can be devided into two areas: quiet regions with relatively weak or no magnetic fields, and active regions with strong concentrations of magnetic fields. The active coronal regions exhibit many loop-shaped structures. Coronal loops are hot and bright and emit a relatively strong X-ray flux. Soft X-ray observations by *Yohkoh* and a high-resolution soft X-ray image (Golub *et al.*, 1992) show that small coronal loops emit X-rays in which rapid variations on time scales of about a minute or less are observable. In some coronal loops isolated patches may suddenly brighten up and when more than two coronal loops interact they can form one continuous bright structure (Shimizu *et al.*, 1992). These observations mean that the energy flux, being transformed in the corona, apparently undergoes rapid changes within small loops.

A solar flare is considered to be a sudden, rapid release of magnetic field energy, stored as a current-system in an active region. It directly heats the corona. Flare events can occur over a wide range of sizes and energies. The smallest flares, the nanoflares are much more numerous than the larger ones. These smallest flares may therefore contribute to the steady heating of the corona (Parker, 1988, 1994).

The energy of the convective motions at the photospheric level can be transformed into other energy forms mainly in two ways: by the formation of MHD waves and of d.c. field-aligned electric currents induced in magnetic loops. MHD waves can be produced by periodic motions in the photosphere and by periodic motions of magnetic flux-tubes, while d.c. field-aligned electric currents are due to relatively slow motions of magnetic flux-tubes (Sakai and Fushiki, 1995). The previously presented theoretical models of coronal heating can therefore be classified into two energy-dissipation mechanisms: dissipation of MHD waves in the corona, or dissipation of d.c. electric currents along the magnetic loops. A fair number of models (cf., Narain and Ulmschneider, 1990, survey paper:1995; also Kuperus *et al.*, 1981) has been proposed for the mechanisms of the energy dissipation. A new current-based heating process in coronal loops to explain the above-mentioned observational features was proposed by Beaufume *et al.* (1992).

It seems important to investigate how a loop-current can be generated near or under the photosphere by different kinds of convective motions. The plasma motions causing current-loops or current-sheet systems can be divided into three types: relative shear motions of different magnetic polarities, collisions of different magnetic polarities, and twisting or rotational motions of magnetic fluxtubes. Here, we concentrate on the third type of plasma motions; twisting or rotational motions of the magnetic fluxtube. We consider two types of evolution of the current-loops: quasi-steady evolution and transient evolution. They are controlled by the time-variations of the plasma vortex motion near the photosphere. It has been shown by a 3-D MHD simulation (Sakai and Fushiki, 1995; *this review* sections 3.1 and 3.2) that transient rotational motions of a magnetic fluxtube near the photosphere can cause a d.c. current with a shell structure, sometimes associated with a return current, depending on the velocity profile of the rotational motion. The simulations shows that current-generation can occur dynamically within a short period of time. Furthermore, *Yohkoh* soft X-ray observation of a flare caused by X-type loop-coalescence (de Jager *et al.*, 1994, 1995; *this review* section 6.4) showed upward moving brightenings along flare loops during a few minutes just before the onset of the flare. This moving transient brightening event may be related to the process of rapid current-dissipation. In the transient current-generation process a loop may start magnetic pinching due to the unbalance of pressure inside and outside the loop. Therefore it is important to investigate the heating process related to transient moving brightenings in loops that may be related to current-dissipation in the loop.

We next discuss how d.c. electric currents along magnetic loops can dissipate rapidly and by what mechanism the background plasma can be heated. We also address the question of explaining the mechanisms of electromagnetic wave emission by a current-carrying loop. By using a three-dimensional electromagnetic particle code, Sakai *et al.* (1994) found that current-carrying loops can be heated by a magnetic pinch driven by the unbalance of pressure inside and outside the loop. Electrons and ions in the loop can both be heated in a direction perpendicular to the ambient magnetic field. As a consequence of the magnetic pinching, the temperature perpendicular to the magnetic field can increase by about a factor of 10 with regard to that along the magnetic field. The temperature anisotropy can induce a plasma instability, by which high-frequency electromagnetic waves can be excited. The plasma current can also excite a kinetic kink instability, which heats ions preferentially in the direction perpendicular to the magnetic field.

In this Section we will therefore meet three kinds of instabilities, which occur successively: (1) a *Buneman instability*, which is driven by a plasma current along a magnetic field; (2) a *pinching instability*, and (3) a *kink instability*. These instabilities will be studied, mainly by simulations based on many-particles codes. In this respect it is useful to make a *general preambulary remark*: Simulation treatments based on a many-particles code, like those described here, present a scaling problem. The simulation loop is very small as compared with a real flare-loop, and so is the number of involved simulation particles. This presents a difficulty, but if the electrons and ions are well magnetized, and if their gyroradii are small as compared with the loop-size, no changes in the basic physics are expected. The advantage of the particle simulation method is that it is an efficient way to study acceleration problems in flare loops.

We present the simulation model and the parameters used in Sakai *et al.*'s (1994) simulation. The simulation used a three-dimensional, fully electromagnetic

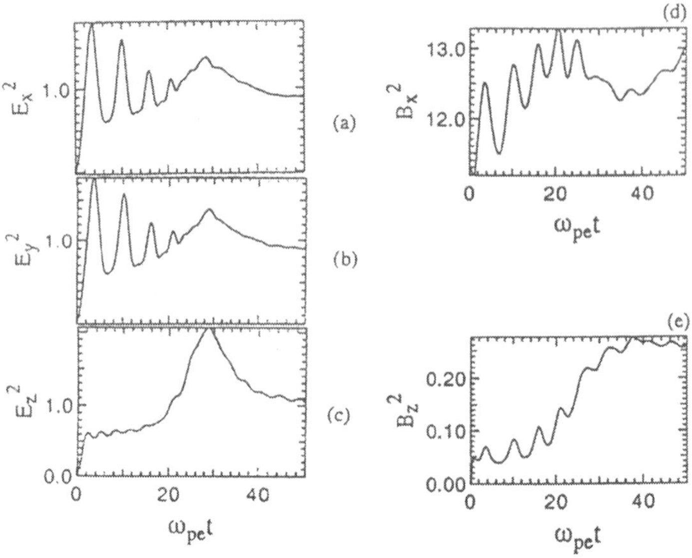

Figure 80. Current-loop heating by pinching magnetic oscilations. Time history of the perturbed field energy in the simulation system; (a) E_x^2, (b) E_y^2, (c) E_z^2, (d) B_x^2, (e) B_z^2. The ordinate scales are arbitrary. (Sakai *et al.*, 1994. From *Solar Phys.*)

and relativistic particle code (see Buneman, 1993, Buneman *et al.*, 1992). The system size used for the simulation is $L_x = L_y = 85\Delta$, $L_z = 160\Delta$ with 2 million particles, where L_x, L_y, and L_z are the lengths of the system in the three dimensions and Δ (=1) is the grid size. A periodic boundary condition is imposed in the direction of the ambient magnetic field (z-direction), while radiating boundary conditions (Lindman, 1975) are imposed in the x- and y-directions.

In the actual simulation described above, the particles are initially distributed homogeneously in the system. The current-carrying electrons are distributed over a column with radius $a = 4.47\Delta$, and the axis of the current-column is located in the center of the domain ($x_{cnt} = y_{cnt} = 43$), directed parallel to the z-axis. The plasma density inside the column is the same as the outside density and the plasma in the column is initially charge-neutral. The electron temperature carrying the current is $0.09T_g$, where T_g is the temperature of the background electrons, which is equal to the ion temperature. The initial electron drift velocity associated with the current is $v_d = 7.0v_{tg}$, where v_{tg} is the background electron thermal velocity. The other parameters are the following: $m_i/m_e = 64$, $\omega_{pe}\Delta t = 0.1$, $\beta = 0.11$, $\lambda_{\rm De} = 0.469\Delta$, $c/\omega_{pe} = 5.0$, $c/v_{tg} = 10.67$, electron gyroradius $\rho_e = 1.17$, ion gyroradius $\rho_i = 9.37$, $V_a/c = 0.05$, $v_{tg} = 0.04688$, ion thermal velocity $v_{it} = 0.005859$. Here, m_i and m_e are the masses of ions and electrons, $\lambda_{\rm De}$ is the Debye-length, and ω_{pe} the electron plasma frequency. The other parameters have been defined already. A poloidal magnetic field, consistent with the current-density J_z, is included.

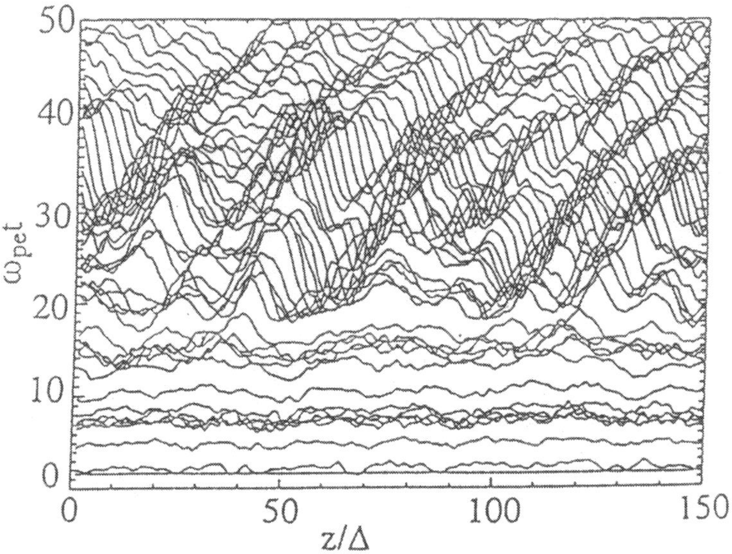

Figure 81. Current-loop heating by pinching magnetic oscillations. Time evolution of the magnetic field component B_x along the z-direction in $x = 43$, $y = 43$, showing the magnetic oscillations. (Sakai *et al.*, 1994. From *Solar Phys.*)

Figure 80 presents the time histories of the electric and magnetic field energies of the simulation system along the three axes. Figure 81 shows the time history of the fluctuations of the x-component of the magnetic field (B_x) in the center of the current-column along the z-direction. On the basis of these calculations Sakai *et al.* (1994) classified the dynamical behavior of the current loops into three phases: Phase 1: ($\omega_{pe}t = 0$–20) shows pinching oscillations; in phase 2 ($\omega_{pe}t = 20$–30) a kinetic kink instability is excited, and in phase 3 ($\omega_{pe}t = 30$–50) ions are heated due to kinetic kink damping.

Figures 80(a–c) demonstrate that the current-loop can undergo pinching due to the pressure unbalance between the regions inside and outside the loop. During the pinching oscillations, inductive electric fields (E_x and E_y) can be driven. These inductive electric fields can heat electrons in the current-column. Figures 82, 83, and 84 show the electron distribution functions in the current-column at three time steps, viz., $\omega_{pe}t = 3$, 6, and 15. It appears from the curve for $\omega_{pe}t = 3$ that the current-carrying electrons can be heated in the x- and y-directions. Around the expansion phase at $\omega_{pe}t = 6$, the electron plasma is slightly cooled down and it is again heated at $\omega_{pe}t = 15$.

During the pinching phase the background ions are also heated in the x- and y-directions, as shown in Figure 85, where the background ion distribution functions are shown for $\omega_{pe}t = 6$ and 15. This current-pinching oscillation can produce a strong temperature anisotropy ($T_\perp \gg T_\parallel$), which may drive another electromagnetic instability (Mikhailovskii, 1975). The electronic temperature anisotropy

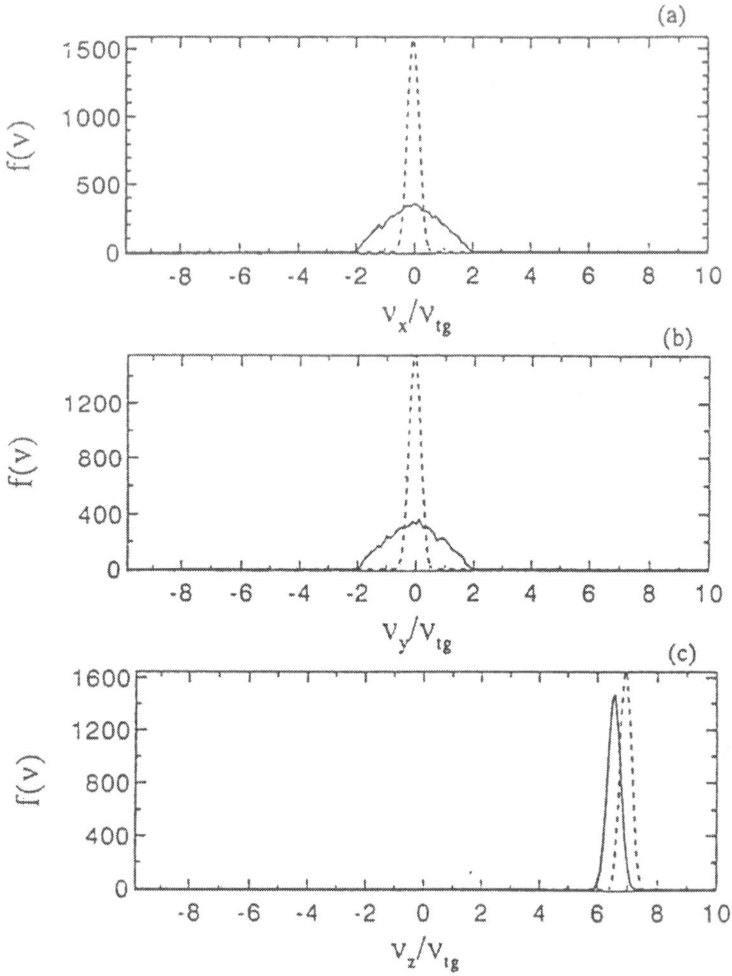

Figure 82. Current-loop heating by pinching magnetic oscillations. Velocity distribution of current-carrying electrons, initially in the column, at $\omega_{pe}t = 0$ (dashed curves) and 3 (solid curves), (a) V_x, (b) V_y, (c) V_z. (Sakai *et al.*, 1994. From *Solar Phys.*)

can excite both low-frequency whistler waves and high-frequency electromagnetic waves. The ionic temperature anisotropy can also excite low-frequency Alfvén waves. In the present simulation the electric current along the magnetic field is also an energy source for exciting a kinetic kink instability.

Figure 86 shows the dispersion relation of the density fluctuations, which can be obtained by performing a two-dimensional Fourier transformation (one coordinate in the z-direction and one in time) for the first 512 time steps. By the same method as in the case shown in Figure 86 the dispersion relation for the magnetic fluctuations (B_x) was obtained. It is shown in Figure 87. It appears from this figure that low-frequency electromagnetic waves are excited in the second phase ($\omega_{pe}t = 20$–30).

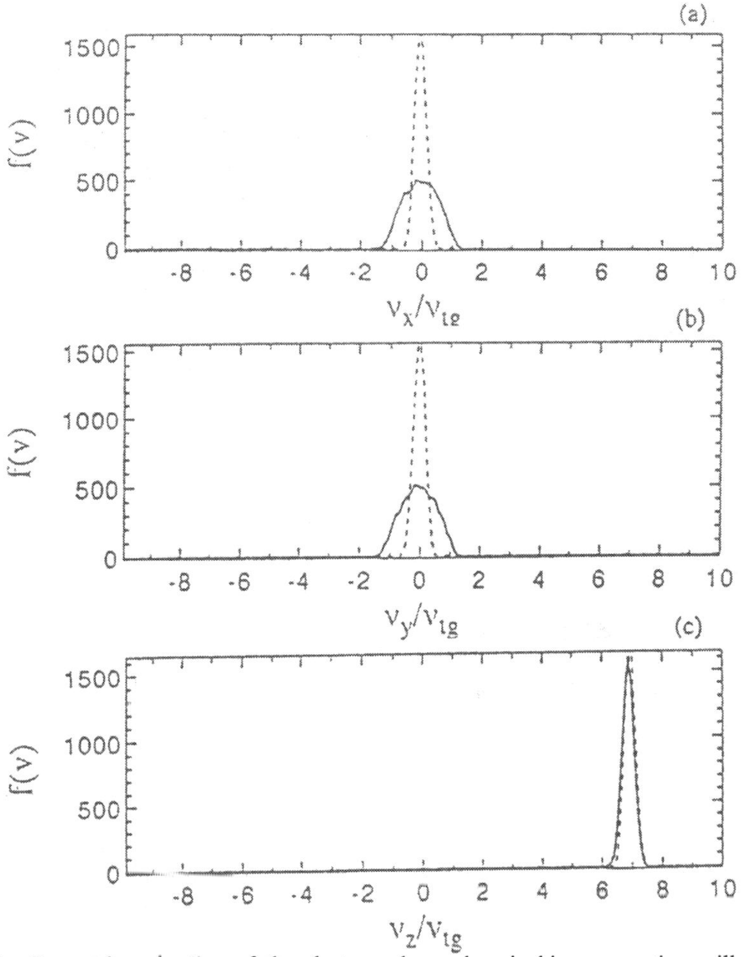

Figure 83. Current-loop heating of the electron plasma by pinching magnetic oscillations. As Figure 82, but for $\omega_{pe}t = 0$ (dashed curves) and 6 (solid curves).

The strong low-frequency waves propagate into the positive z-direction with the same velocity as the strong density perturbations, as can be seen in Figure 86. That figure shows also that high-frequency electron plasma waves are excited in the current-column. These strong low-frequency waves have a right-hand polarization signature. Hence, a helical perturbation, right-hand polarized, can be driven by the kinetic kink instability. The mode number of this instability, calculated for $krB_z(r)/B_\theta(r) = 1$, $k = 2\pi n/L$ ($r = a, n = 8$), shows good agreement with the simulation results. In the simulation the Alfvén velocity is $V_A/c = 0.05$. An Alfvén transit time of $\tau_A = a/V_A = 16\omega_{pe}^{-1}$, is obtained by using $a = 4\Delta$. Therefore, the growth time of the kinetic kink instability obtained in the simulation is consistent with this Alfvén transit time.

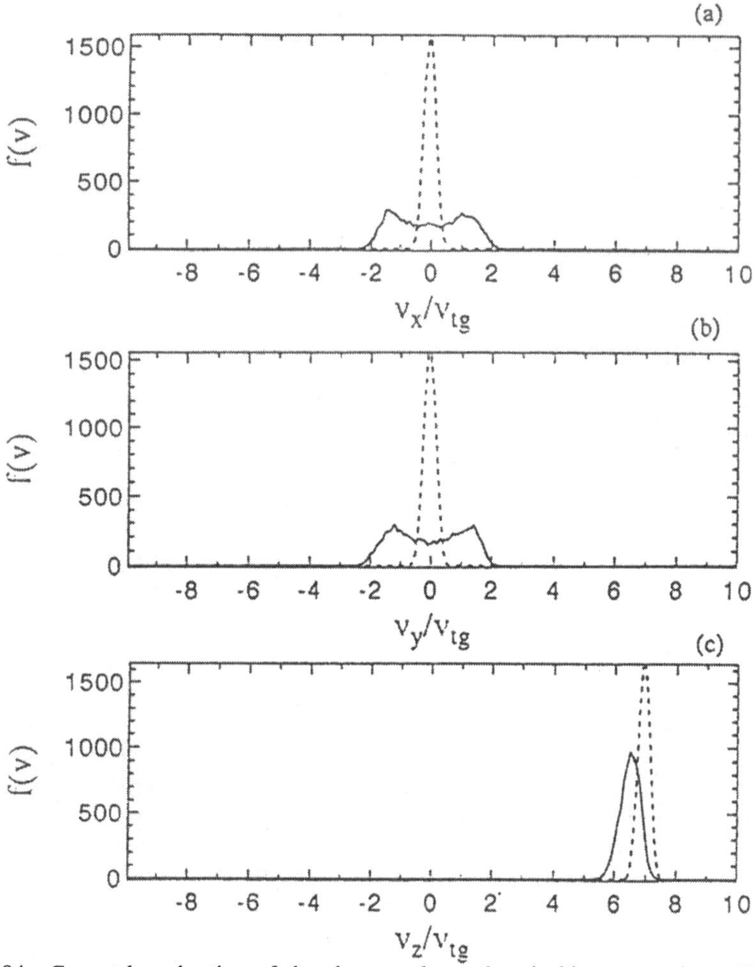

Figure 84. Current-loop heating of the electron plasma by pinching magnetic oscillations. As Figure 82, but for $\omega_{pe}t = 0$ (dashed curves) and 15 (solid curves).

Figure 88 presents the dispersion relation for the magnetic field fluctuations (B_x) for a line outside the current column ($x = 43, y = 20$). Sakai *et al.* (1994) found that high-frequency electromagnetic waves are excited and that they propagate away from the current-loop. These electromagnetic waves can be excited strongly in the direction of the electron drift. They show fundamental and second harmonics. It appears from Figure 88 that low-frequency whistler waves also propagate away from the current-loop.

It has already been demonstrated by the results in Figure 81, that large-amplitude kinetic kink waves, excited during phase 2 (defined, as above, as the period between $\omega_{pe}t = 20$) and 30, are strongly damped after $\omega_{pe}t = 40$. The ion distribution functions at $\omega_{pe}t = 30$ and 51 are shown in Figures 89 and 90.

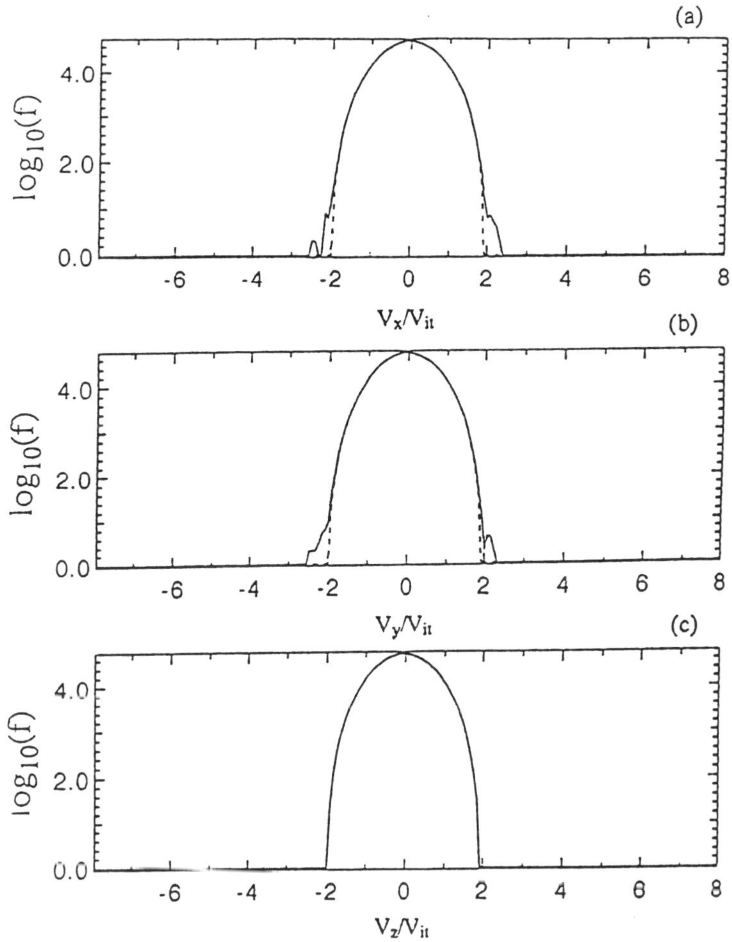

Figure 85. Current-loop heating of the *ion* plasma by pinching magnetic oscillations. As Figure 82, for $\omega_{pe}t = 0$ (dashed curves) and 15 (solid curves).

By comparing the ion distribution functions at the two time steps ($\omega_{pe}t = 30$ and 51), it was found that ions can be heated even more after $\omega_{pe}t = 30$, but only in the x- and y-directions. On the other hand, electrons can not be heated during this period, as seen in Figures 91 and 92, where the electron distribution functions at $\omega_{pe}t = 30$ and 51 are shown. We conclude therefore that in the third phase ions are accelerated in a direction perpendicular to the magnetic field by the large amplitude kinetic kink waves.

As was already demonstrated in Figure 86, strong electrostatic fields associated with the density hump due to kinetic kink waves are excited. They propagate with the same velocity as the kinetic kink wave. Here we (Sakai *et al.*, 1994) show, from the calculation of a test ion orbit in the right-hand polarized kinetic kink waves, moving in a strong electrostatic field, that ions trapped in a strong potential well

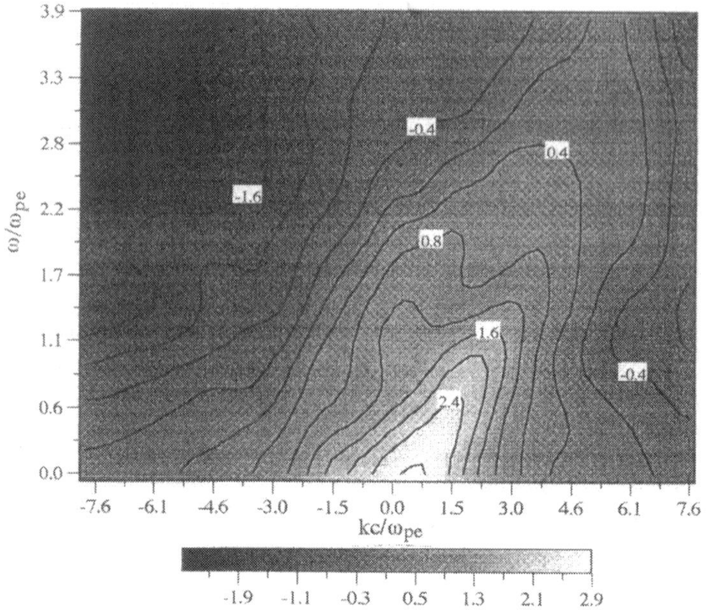

Figure 86. Dispersion relation for the charge density in a heated loop. Wave intensity as a function of ω and k_z for $22 \leq z \leq 150$. The values of the charge density used were taken from the calculated time-dependent charge-density data, as given in Figure 81. (Sakai *et al.*, 1994. From *Solar Phys.*)

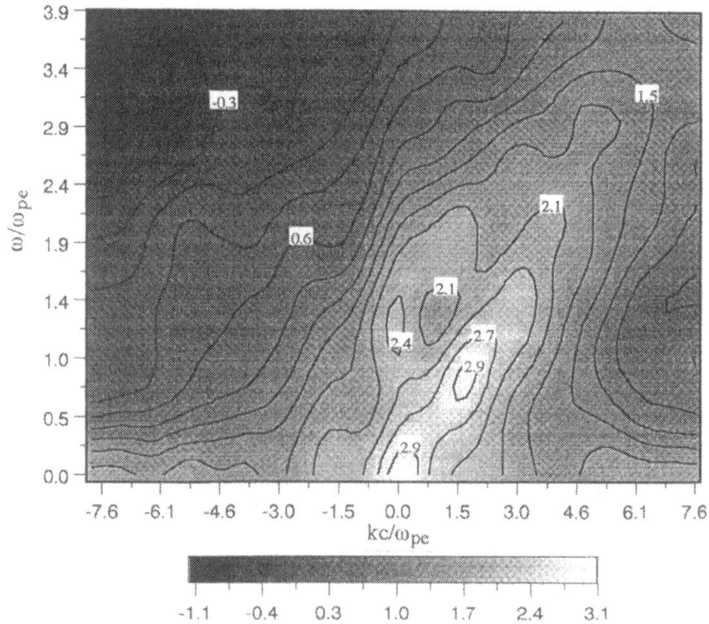

Figure 87. Dispersion relation for B_x in $x = 43$ and $y = 43$ in a heated loop. Other data are as in Figure 86.

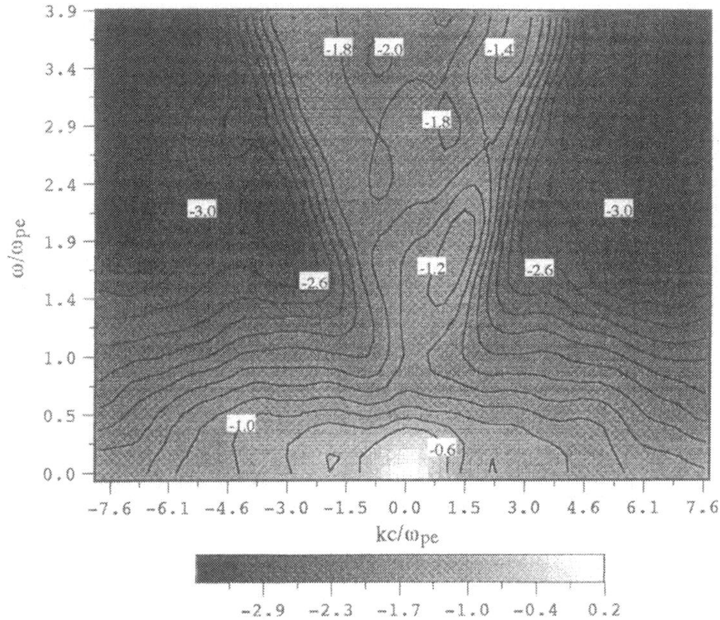

Figure 88. Dispersion relation for B_x in $x = 43$ and $y = 20$ in a heated loop. Other data are as in Figure 86.

can be accelerated. The equation of motion of the test ion is given in coordinates moving with the phase velocity of the kinetic kink wave, as follows:

$$\frac{\mathrm{d}v_\perp}{\mathrm{d}t} = -\frac{B_\perp}{k\rho_i}\frac{\mathrm{d}\xi}{\mathrm{d}t}\sin(\xi - \phi)$$

$$\frac{\mathrm{d}\phi}{\mathrm{d}t} = -\frac{m_e}{m_i}\frac{\omega_{ce}}{\omega_{pe}} + \frac{B_\perp}{k\rho_i v_\perp}\frac{\mathrm{d}\xi}{\mathrm{d}t}\sin(\xi - \phi)$$

$$\frac{\mathrm{d}^2\xi}{\mathrm{d}t^2} = E_1\sin(\xi) + \left(\frac{m_e\omega_{ce}}{m_i\omega_{pe}}\right)^2 k\rho_i B_\perp \sin(\xi - \phi)$$

where

$$\xi = kz - \omega t,\ E_1 = \omega_b/\omega_{pe}{}^2, \quad \omega_b = (ekE_z/m_i)^{1/2},$$

and

$$B_\perp = (B_x{}^2 + B_y{}^2)^{1/2}/B_{z0}.$$

The time in the above equations is normalized by ω_{pe}^{-1} and $v_\perp$ is normalized by the ionic thermal velocity. If the first term at the right-hand side of the third

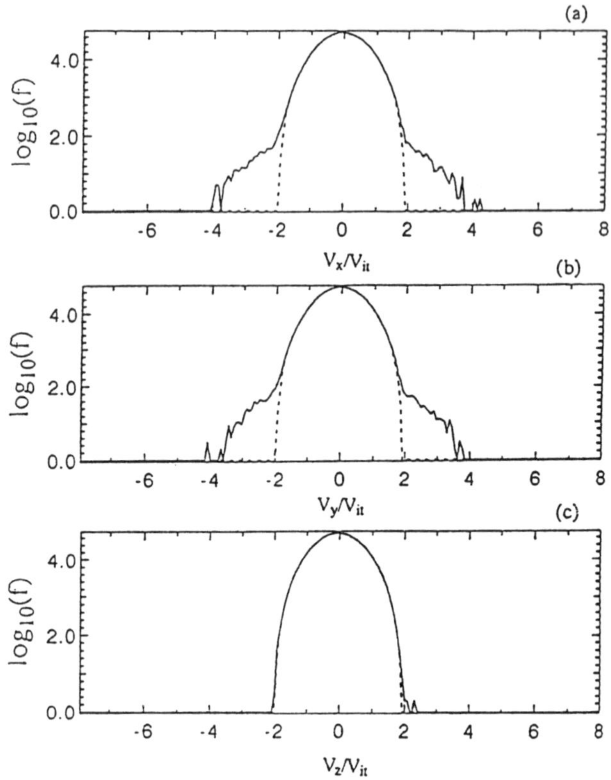

Figure 89. Ion acceleration in a heated loop. Velocity distribution of ions in the simulation system at at $\omega_{pe}t = 0$ (dashed curves) and 30 (solid curves), (a) V_x, (b) V_y, (c) V_z. (Sakai *et al.*, 1994. From *Solar Phys.*)

equation exceeds the second term, ions can be trapped in an electrostatic potential well. The trapped ions are accelerated in a direction perpendicular to the magnetic field. Figure 93 shows results obtained from numerical calculations with the above equations. The maximum velocity, obtainable by this mechanism, is approximately $(B_\perp/k\rho_i)v_{it}$. Therefore ion acceleration becomes effective when the kinetic kink wave amplitude is large enough and when the wavelength of the excited kinetic kink waves is larger than the ionic gyroradius.

Such processes are of importance for understanding the mechanism of coronal loop heating, especially for explaining the rapid and small-scale heating processes observed by high-resolution soft X-ray images (Golub *et al.*, 1990; Inda-Koide, 1994). The results obtained in the simulation, viz., that high-frequency electromagnetic waves are emitted by a heated loop, should be checked by simultaneous observations in radio and soft X-ray images.

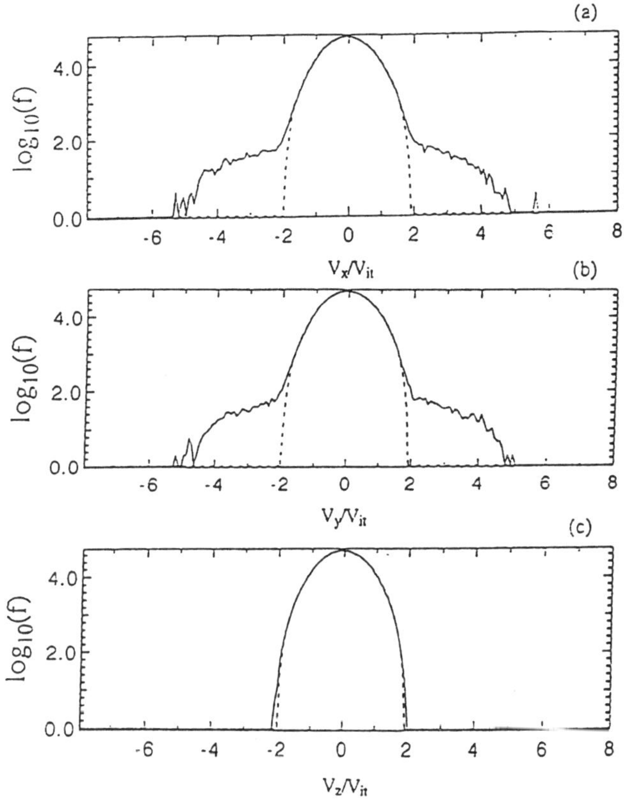

Figure 90. Ion acceleration in a heated loop. As Figure 89, but for $\omega_{pe}t = 0$ (dashed curves) and 51 (solid curves).

5.2. Coalescence Dynamics of Kink-Unstable Loops

Yohkoh observations suggest that studying the three-dimensional aspects of the processes of loop-collisions may be important for understanding transient coronal loop heating as well as high-energy particle acceleration during the impulsive phase of solar flares. For investigating these processes, it was necessary to develop a three-dimensional electromagnetic particle code. Simulation results by Nishikawa *et al.* (1994) show that prior to two-loops coalescence, each loop first shows a pinching oscillation and thereafter becomes unstable with helical perturbations. The dynamical behavior of a single current-loop has been analyzed by Sakai *et al.* (1994). During the pinching oscillations electron heating perpendicular to the ambient magnetic field becomes apparent. Just after coalescence of two helical loops a tilting motion is induced and finally the helical structure relaxes, which results in strong electron- and ion-heating, mostly in a direction perpendicular to that of the initial current.

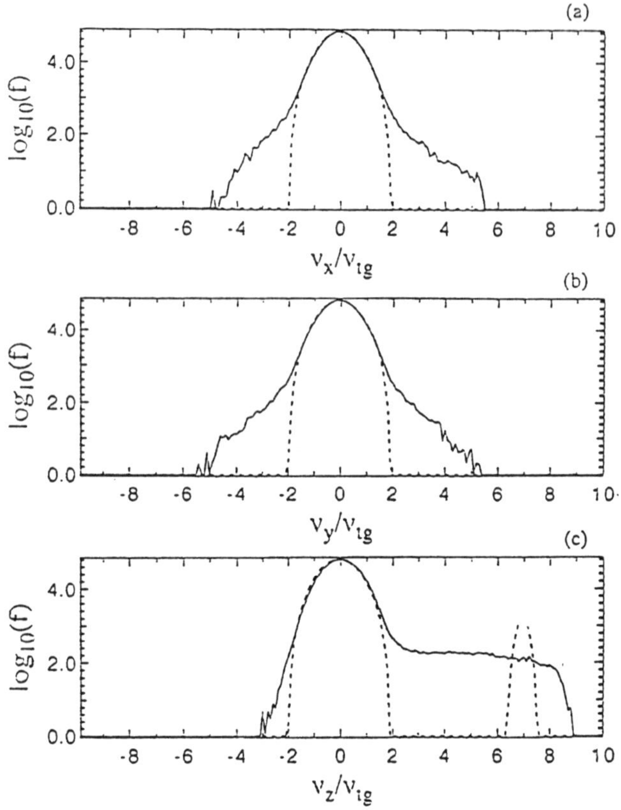

Figure 91. Electron acceleration in a heated loop. Velocity distribution for the electrons. Further caption as in Figure 89.

The electrons carrying the currents in the two loops are initially in two columns, oriented along the z-direction. In the simulations the loops have a radius $a = 4\Delta$, and the two centers of them are located at $x_{c1} = 32, y_{c1} = 43$, and $x_{c2} = 52, y_{c2} = 43$. The plasma density inside the current-column is the same as outside it. The numbers of current-electrons within the columns are 7780 and 7789, respectively. The electron drift-velocity of both currents is the same, $V_d = 7.0v_{et} (= 0.66c)$, where v_{et} is the electron thermal velocity and $c = 0.5$. The electron temperature associated with the currents is $T = 0.09T_b$, where T_b is the temperature of the background electrons. The following parameter values were chosen: $m_i/m_e = 64$, $T_i/T_e = 1, \omega_e/\omega_{pe} = 0.4, c/v_{et} = 10.67, \omega_{pe}\Delta t = 0.1, \beta = 0.11, \lambda_{\rm De} = 0.469\Delta$, $\rho_e = 1.17\Delta, \rho_i = 9.38\Delta, c/\omega_{pe} = 5.0\Delta$. (The symbols are the same as those in the similar study described in section 5.1.) The ambient magnetic field is chosen to be $B_{zo} = 0.1$. Quiescent initial conditions are achieved by including an initial poloidal magnetic field consistent with the current density J_z carried by the electrons in the two columns at $\omega_{pe}t = 0$ (Nishikawa *et al.*, 1994). The maximum value of the

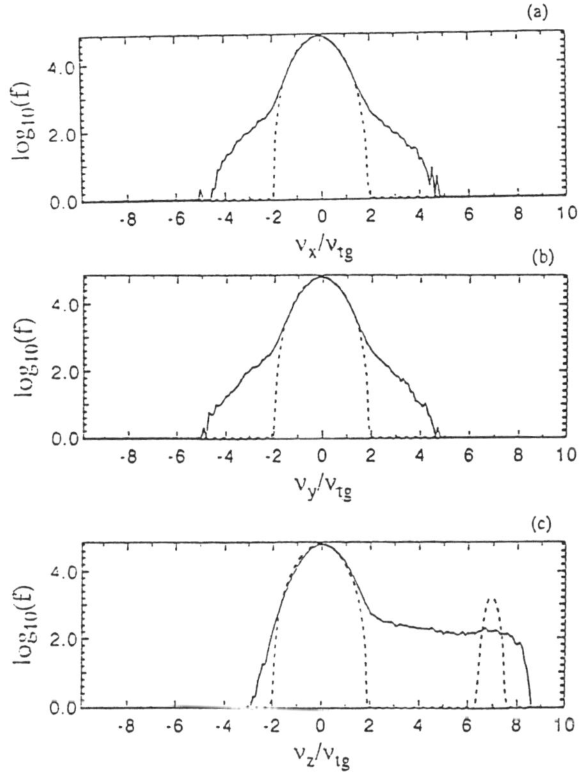

Figure 92. Electron acceleration in a heated loop. Velocity distribution for the electrons. Further caption as in Figure 90.

ratio between the poloidal and ambient magnetic fields at the edge of the column is initially $B_\theta/B_{zo} = 0.8$.

Figure 94 shows the time history of the current-carrying electrons in a three-dimensional system. The background particles are not drawn. Based on such studies, Nishikawa *et al.* (1994) classified the dynamical behavior of the process of coalescence of two current-loops into three phases: (1) pinching oscillation of each current-loop, (2) excitation of a kinetic kink instability with helical perturbations, and (3) coalescence and relaxation. As seen from Figure 94 and from the time history of the electric field energies in Figure 95, each current-loop undergoes pinching oscillations due to the unbalance between the pressure inside and outside the loops, until $\omega_{pe}t = 20$. In the second phase, following the pinching oscillation, a dominant kinetic kink instability is excited. Helical, right-hand polarized perturbations are driven by this instability. The mode number, calculated by $krB_z(r)/B_\theta(r) = 1$, $k = 2\pi n/L$ $(r = a, n = 8)$ agrees well with the simulation results. In the simulation the Alfvén velocity is $V_{\rm A}/c = (m_e/m_i)^{1/2}\omega_{ce}/\omega_{pe} = 0.05$. We can get the Alfvén transit time, $\tau_{\rm A} = a/V_{\rm A} = 16\omega_{pe}^{-1}$, by using $a = 4\Delta$. Therefore, the

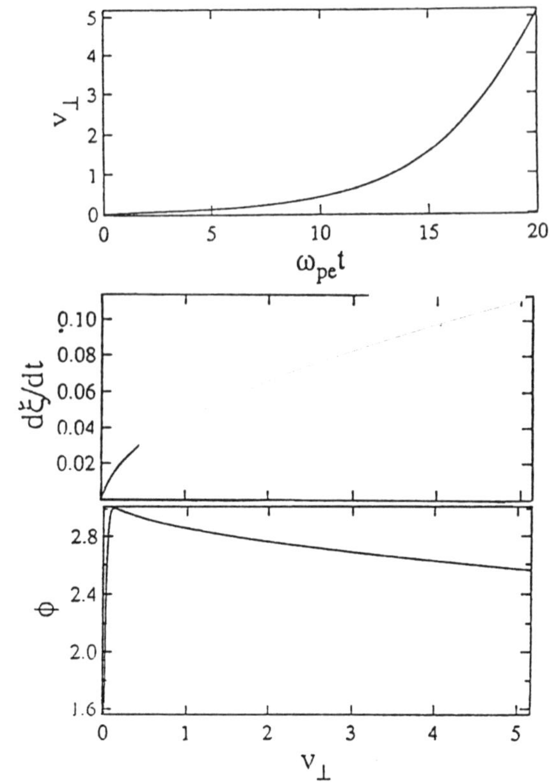

Figure 93. Acceleration of a test ion. *Above*: time history of the velocity component perpendicular to **B** of a test ion trapped in the electric potential. *Middle* and *lower* panel: orbit of the ion in velocity space. (Sakai *et al.*, 1994. From *Solar Phys.*)

growth-time of the kinetic kink instability observed in the simulation is consistent with the Alfvén transit time.

On the other hand, it is also possible to observe a Buneman (beam) instability. For V_d-values around $V_d = 3.5v_{et}$ a beam instability is dominantly excited and bunched electrons, due to this instability, excite whistler waves, as has been studied previously (Omura and Matsumoto, 1988; Pritchett *et al.*, 1989; Nishikawa *et al.*, 1993). These different generation mechanisms of whistler waves were studied by Sakai *et al.* (1994). The two loops with strong right-hand polarized helical perturbations begin to collide in the third phase. Just after the loop-coalescence, a tilting motion with left-hand rotation along the z-axis is induced and the helical pattern starts to propagate along the z-axis. After the loop-coalescence the helical perturbations relax into a single thick loop, in which the initial currents nearly fully decay.

Figure 95 gives the time-history of the three components of the electric field energies as well as the magnetic field energies. Before the two loops approach each

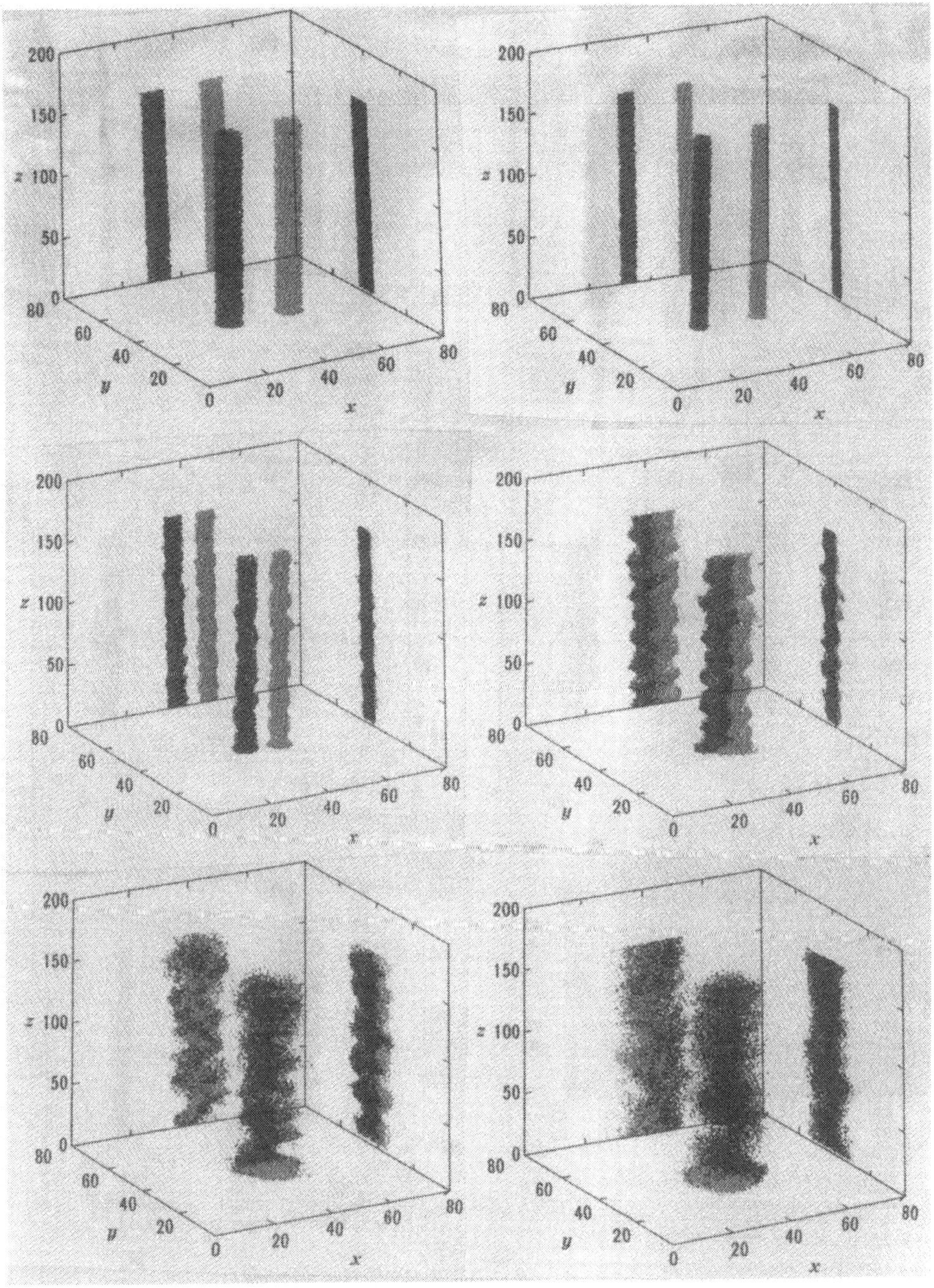

Figure 94. Particle simulation of two-loops coalescence. The various diagrams show the behaviour of the two columns (simulating interacting loops) with time. Time history of the event: From *left* to *right* and from *above* to *below*: $\omega_{pe}t = 0, 3, 15, 21, 33,$ and 48. (Nishikawa *et al.*, 1994. From *Astrophys. J.*)

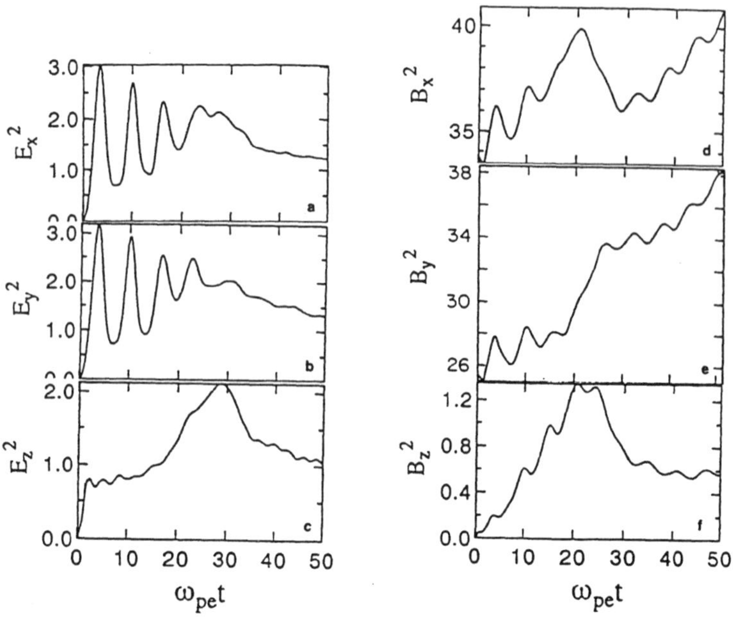

Figure 95. Particle simulation of two-loops coalescence. Time history of the perturbed field energy in the simulation system: (a) $E_x{}^2$; (b) $E_y{}^2$; (c) $E_z{}^2$; (d) $B_x{}^2$; (e) $B_y{}^2$; (f) $B_z{}^2$. (Nishikawa *et al.*, 1994. From *Astrophys. J.*)

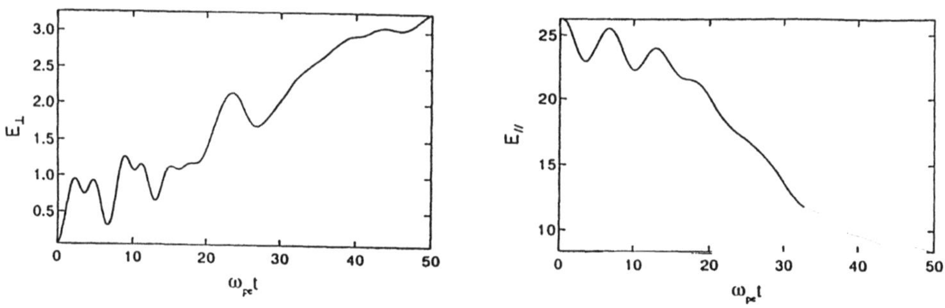

Figure 96. Particle simulation of two-loops coalescence. Time history of the kinetic energy of current-carrying electrons in the columns: (a) perpendicular, (b) parallel. (Nishikawa *et al.*, 1994. From *Astrophys. J.*)

other, at about $\omega_{pe}t = 18$, the two current-loops repeatedly pinch and expand. This is seen in the oscillations of the electric field energy E_x and E_y in Figure 95.

Figure 96 presents the time history of the electron kinetic energies in the current-columns. The electron kinetic energy parallel to the ambient magnetic field decreases, while the electron kinetic energy perpendicular to the magnetic field increases, at the same time as the loop-oscillations. During the development of the kinetic kink instability, the electrons in the simulation system are heated strongly, as seen

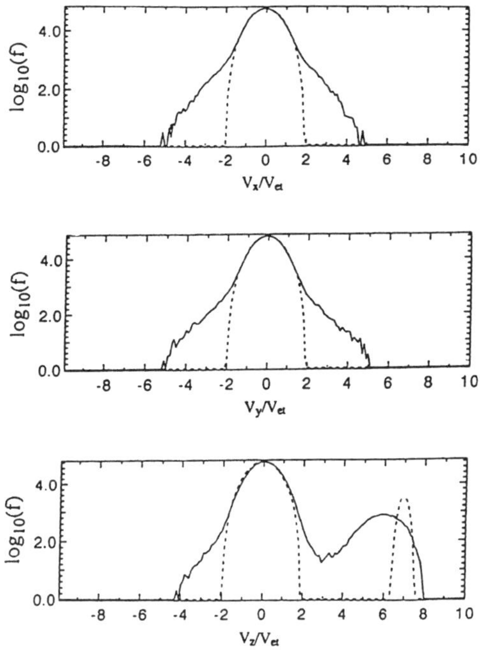

Figure 97. Particle simulation of electrons during two-loops coalescence. Velocity distribution of the electrons in the simulation system at $\omega_{pe}t = 0$ (dashed curves) and 21 (solid curves). *Above*: V_x, *middle*: V_y, *below*: V_z. (Nishikawa et al., 1994. From *Astrophys. J.*)

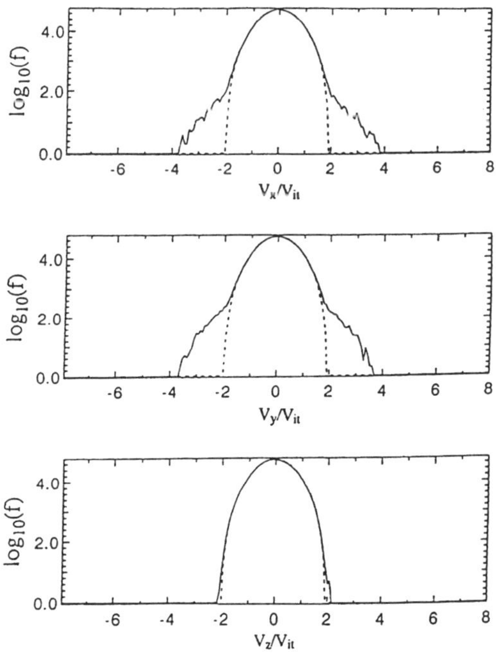

Figure 98. Particle simulation of ions during two-loops coalescence. Caption as in Figure 97, but here for the ions.

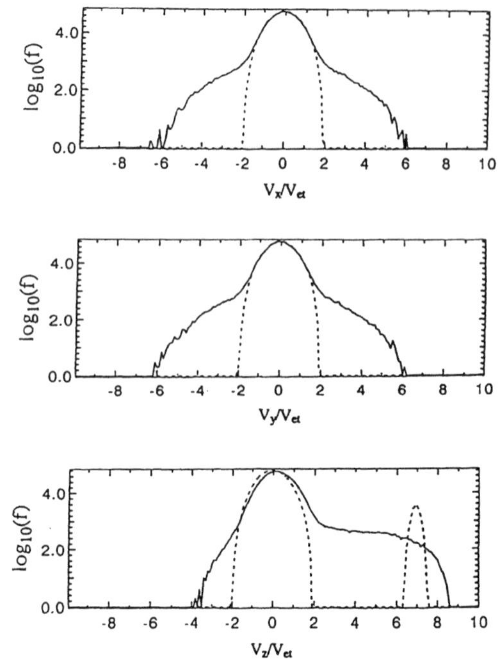

Figure 99. Particle simulation for electrons in two-loops coalescence. Caption as in Figure 97, but for electrons at $\omega_{pe}t = 0$ (dashed curves) and 48 (solid curves).

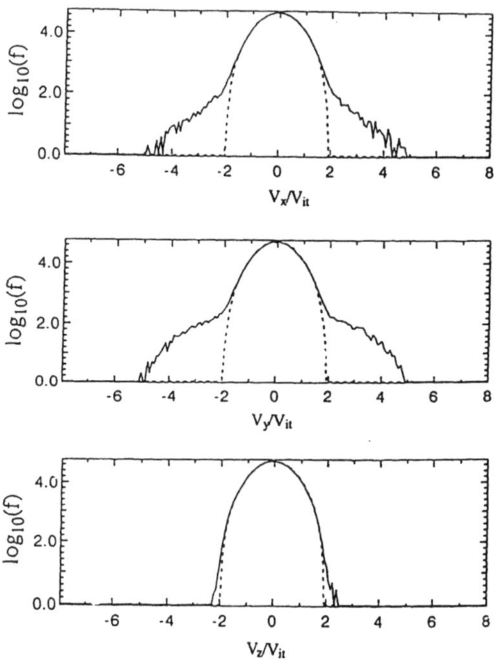

Figure 100. Particle simulation for ions in two-loops coalescence. Caption as in Figure 97, but for ions at $\omega_{pe}t = 0$ (dashed curves) and 48 (solid curves).

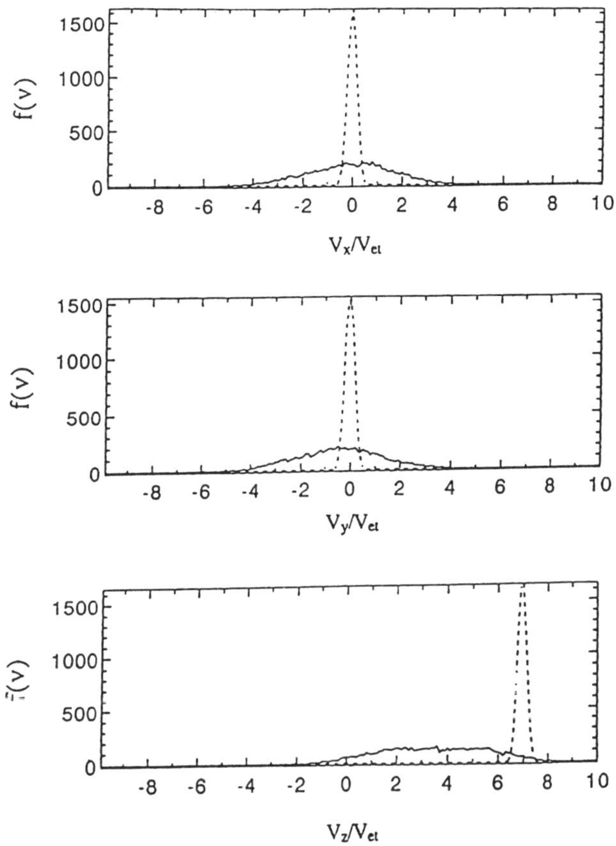

Figure 101. Particle simulation of electrons in two-loops coalescence. Velocity distribution of current-carrying electrons in the column at $\omega_{pe}t = 0$ (dashed curves) and 48 (solid curves). *Above*: V_x, *middle*: V_y, *below*: V_z. (Nishikawa *et al.*, 1994. From *Astrophys. J.*)

in Figure 97. The ions are only heated weakly as compared with the electrons, as shown by Figure 98.

The plasma heating can be strongly enhanced after the coalescence of the two helical loops. Figures 99 and 100 show the velocity distribution functions, both of ions and of electrons in the whole system after the coalescence. Electrons and ions can both be heated by about a factor 9 with respect to the initial temperature. Figure 101 shows the electron distribution functions in one of the current-loops at $\omega_{pe}t = 48$. It appears from the Figure that the electrons can be heated by about a factor 50 with respect to the initial temperature. Heating is mainly in the direction perpendicular to the magnetic field.

Figure 102 contains the time history of the wave patterns of the magnetic fluctuations B_x, which are observed along the z-direction in the point where the two loops coalesced ($x = y = 43$). The dispersion relation of the magnetic perturbations is shown in Figure 103. It was obtained by performing a two-dimensional

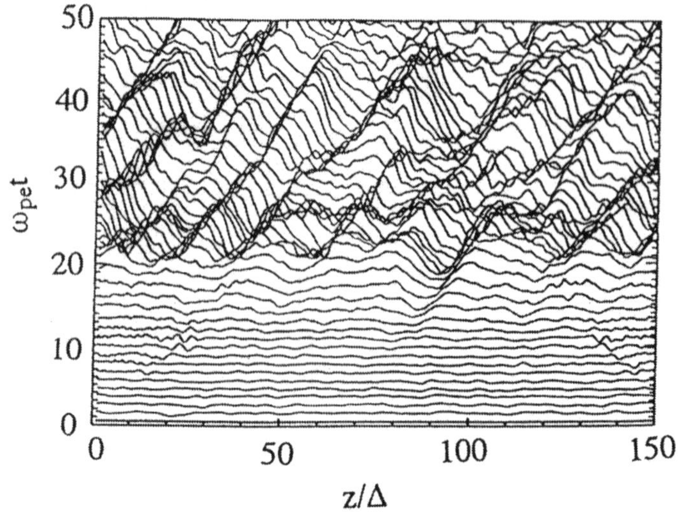

Figure 102. Particle simulation of two-loops coalescence. Time evolution of the magnetic field component B_x along the z-direction at $x = 43$, $y = 43$. Time and distance are normalized as indicated. (Nishikawa *et al.*, 1994. From *Astrophys. J.*)

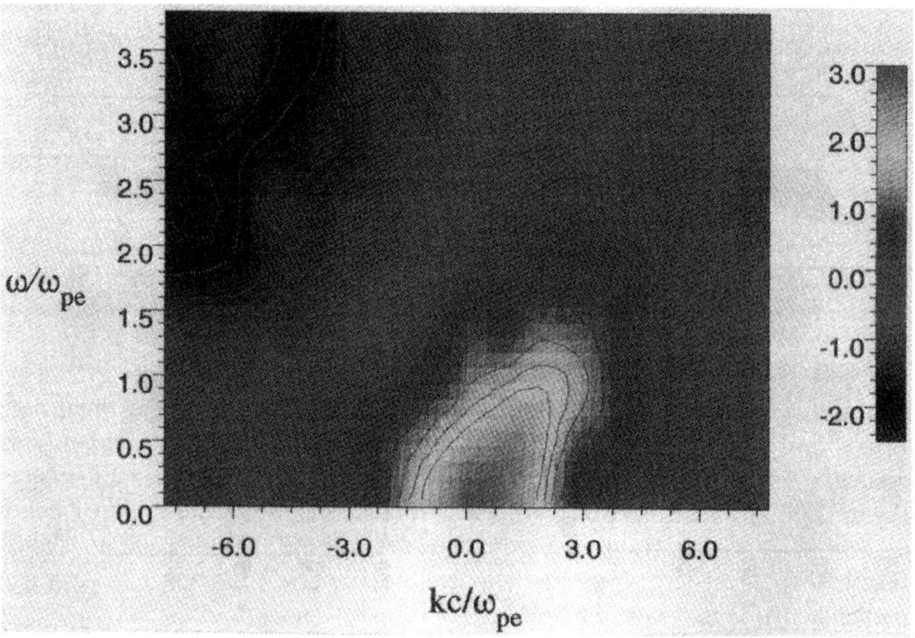

Figure 103. Particle simulation of two-loops coalescence. Wave intensity as a function of the normalized values of ω and k_z along $22 \leq z \leq 150$ for the B_x-values as given in Figure 102. (Nishikawa *et al.*, 1994. From *Astrophys. J.*)

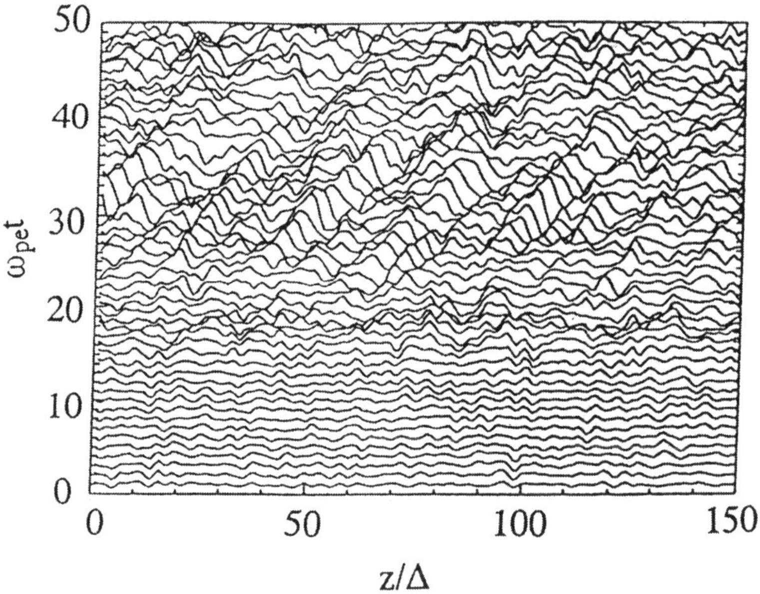

Figure 104. Particle simulation of two-loops coalescence. Time evolution of the charge density along the z-direction at $x = 43$, $y = 43$. (Nishikawa *et al.*, 1994. From *Astrophys. J.*)

Fourier transformation (one space- and one time-coordinate) for the first 512 time steps. The phase velocity of the waves is calculated from Figures 102 and 103 and it is approximately $3v_{et}$. Figure 104 presents the time history of the wave patterns of the density perturbations, which are observed along the z-direction in the same region as the above-described magnetic fluctuations.

Figure 105 gives the dispersion relation of the density fluctuations; the data were obtained in the same way as those in Figure 103. As was already obvious in Figure 103, large-amplitude low-frequency magnetic waves associated with density perturbations propagate along the coalesced loop. The dominant frequency is close to the electron-cyclotron frequency. The large-amplitude helical perturbations with right-hand polarization, which are produced by d.c. electric currents heat ions mostly in a direction perpendicular to the magnetic field. After the coalescence of the two loops, the loops are tilted with a left-hand rotation around the z-axis, while the helical perturbations can be enhanced and lead to further ion heating.

Nishikawa *et al.* (1994) found, from the dispersion relation obtained outside the coalesced loop, that both low-frequency waves as well as high-frequency electromagnetic waves can be generated from the coalesced current-loop. In order to obtain further insight in this radiation, its intensity was compared with that for the case without a current. Clearly, current-loop coalescence leads to the emission of high frequency waves, with an intensity at least by a factor three above the thermal noise level. The high-frequency electromagnetic waves can be excited strongly in the direction of the electron drift velocity. Further studies of the excitation

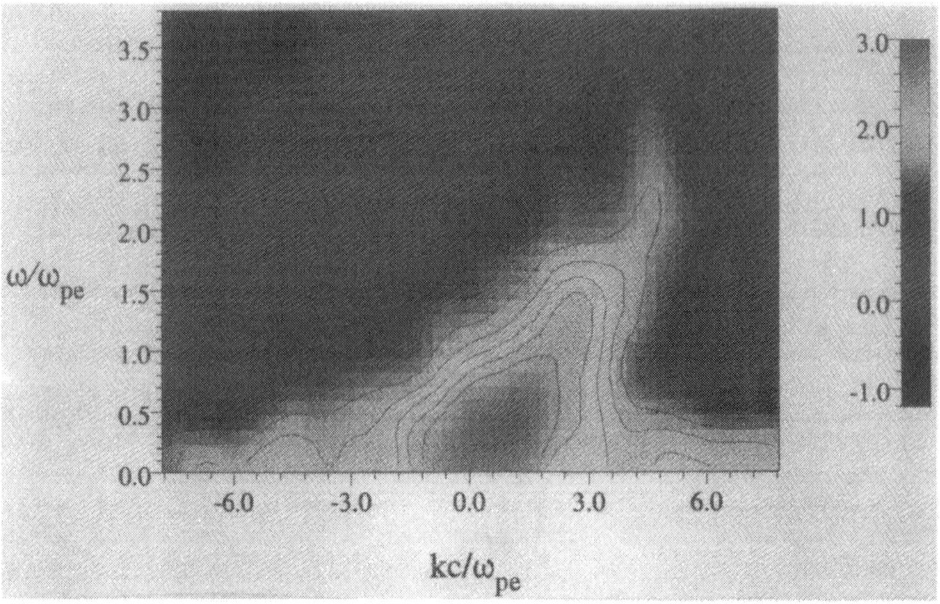

Figure 105. Particle simulation of two-loops coalescence. Wave intensity as a function of ω and k_z along $22 \leq z \leq 150$ for the time variation of the charge density as given in Figure 104. (Nishikawa *et al.*, 1994. From *Astrophys. J.*)

mechanism of the high-frequency electromagnetic waves are in progress (Zhao *et al.*, 1995a). A temperature anisotropy can be produced by pinching oscillations of the currents and is further enhanced after the two-loops coalescence. Therefore, the observation of enhanced electromagnetic wave emission, from the region where two current-loops collide, will be a good signature and a good indicator of such loop interactions.

5.3. High-Energy Electron Acceleration during Two-Loops Coalescence

In section 4.5 we discussed high-energy particle acceleration during coalescence of two loops. In section 5.2 we showed the results of coalescence of two current-loops, which are unstable against a kinetic kink instability. The radius of the current-loop used in the simulation was small as compared with the ion Larmor radius. In this section we follow a study by Zhao *et al.* (1995b), which was based on an assumed larger radius of the loop, being about 2.5 times the ion Larmor radius. The study aimed at investigating the partial reconnection process during two-loops coalescence, by using a model in which the $x - y$ system size ($L_x = 200\Delta$, $L_y = 160\Delta$, $L_z = 40\Delta$) is larger than that used in the model of Nishikawa *et al.* (1994) as described in the previous section, while the system size in the z-direction is reduced in this new study. In this investigation, the kinetic kink instability

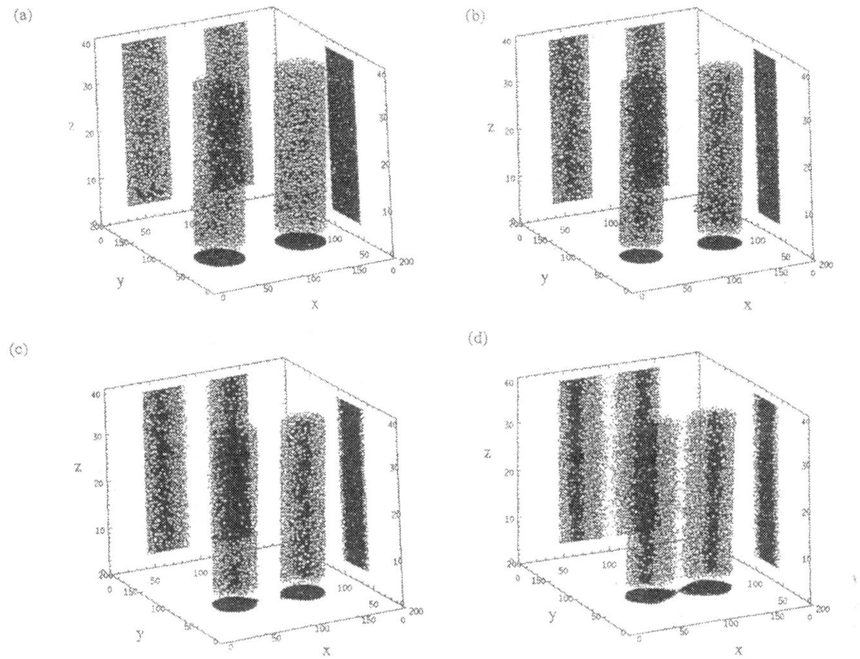

Figure 106. Particle simulation of two-loops coalescence. Three-dimensional development of the positions of the current-carrying electrons, initially in the two columns, with the projections on the xy-, xz-, and yz-planes, at $\omega_{pe}t$ = (a) 0, (b) 9, (c) 15, (d) 21. (Zhao *et al.*, 1995b.)

observed in the previous section, does not appear, due to the short system length in the z-direction.

Figure 106 shows the time development of the spatial distribution of the electrons that carry the two currents, while the background electrons and ions are not shown here. As seen in the figure, the electrons located near the interaction region of the two loops undergo stronger mutual attraction than electrons located further away, which is due to the inhomogeous character of the attraction. This causes an inhomogeneity in the electron density of the coalescing two loops.

Figure 107 presents the time history of the electrons that were initially located in the central region between the two loops. It appears that the electron gas is compressed in the x-direction near the induced current-sheet and it streams out into the y-direction due to magnetic reconnection.

Figures 108 and 109 give the time history of the magnetic fields in the $z = 20\Delta$ plane. It appears that a tilting motion occurs around the magnetic X-point after coalescence. During this tilting motion the magnetic reconnection still continues and the electrons are accelerated as is shown in the subsequent figures.

Figures 110–115 demonstrate that ions and electrons in the central region between the two loops are both heated, mostly by adiabatic compression, while the

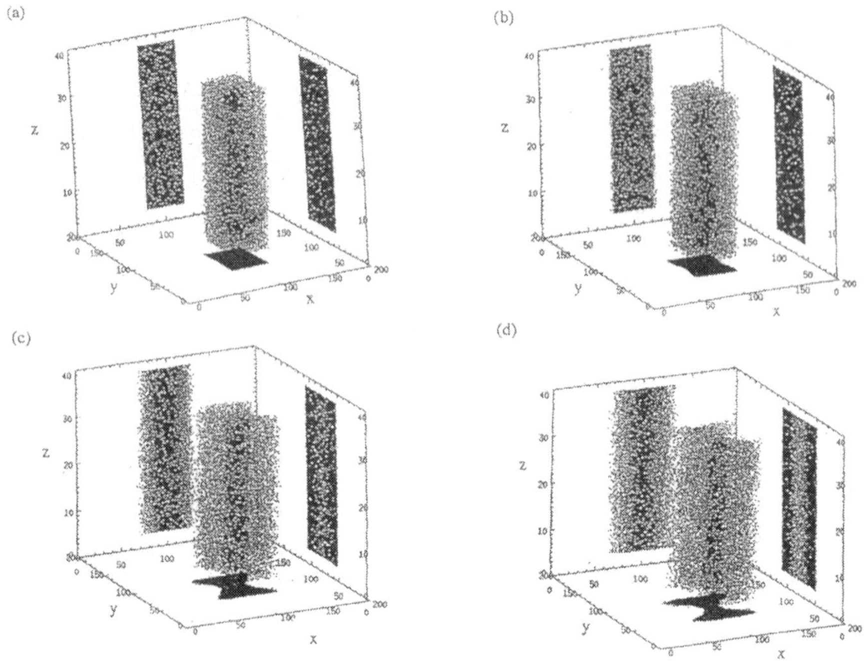

Figure 107. Particle simulation of two-loops coalescence. The same plot as the one shown in Figure 106 but for the electrons that were initially in the region of interaction. (Zhao *et al.*, 1995b.)

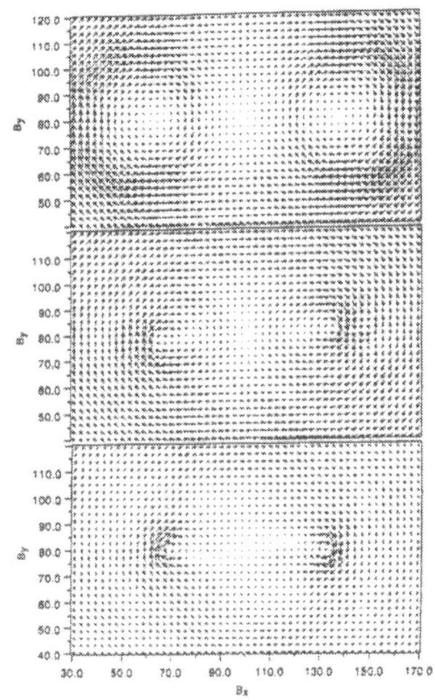

Figure 108. Particle simulation of two-loops coalescence. Time history of (B_x, B_y) vector plots in the plane $z = 20$ at $\omega_{pe}t = 0$ (*above*), 20 (*middle*), and 30 (*below*). (Zhao *et al.*, 1995b.)

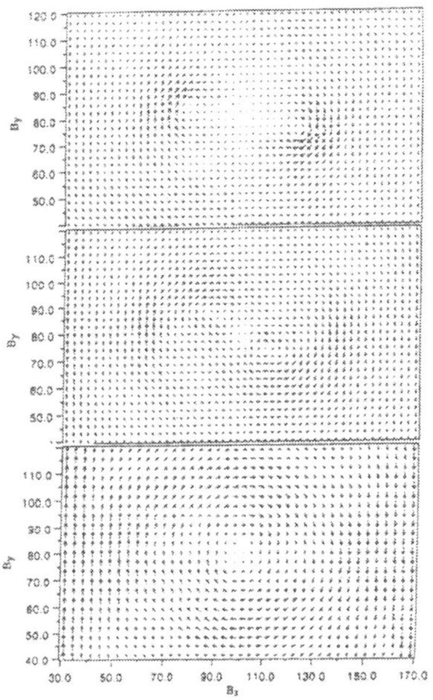

Figure 109. Particle simulation of two-loops coalescence. As Figure 108, but for $\omega_{pe}t = 45$ (*above*), 60 (*middle*), and 75 (*below*). (Zhao *et al.*, 1995b.)

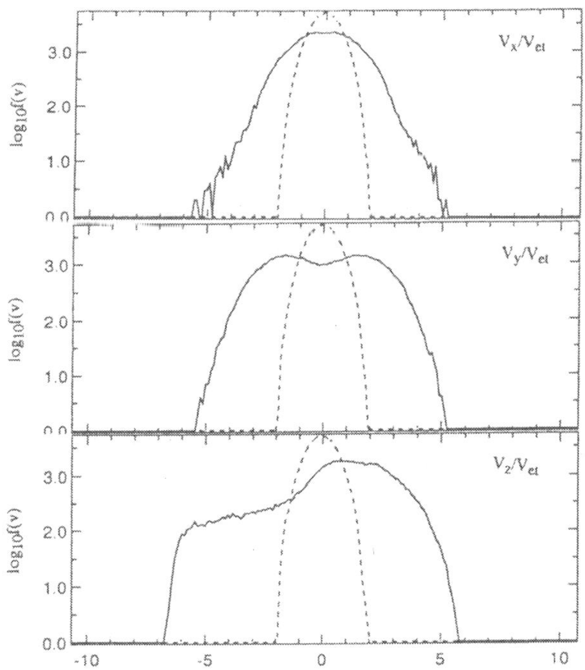

Figure 110. Particle simulation of two-loops coalescence. The three components of the velocity distribution of the electrons in the interaction region at $\omega_{pe}t = 20$. (Zhao *et al.*, 1995b.)

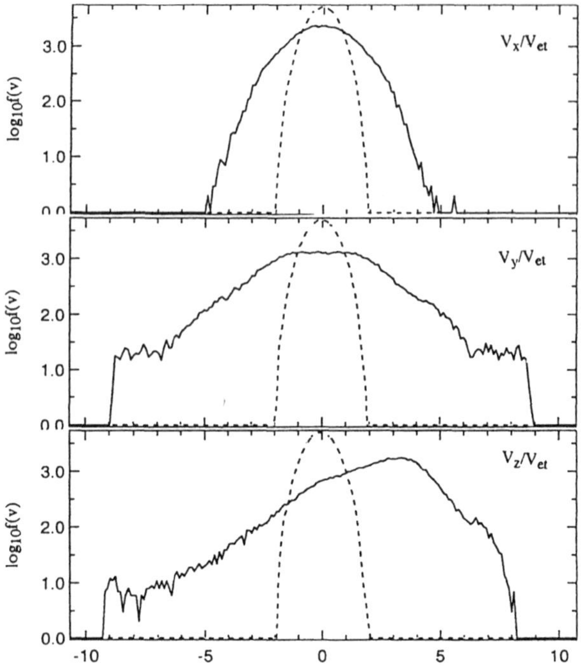

Figure 111. Particle simulation of two-loops coalescence. As Figure 110, but for $\omega_{pe}t = 30$.

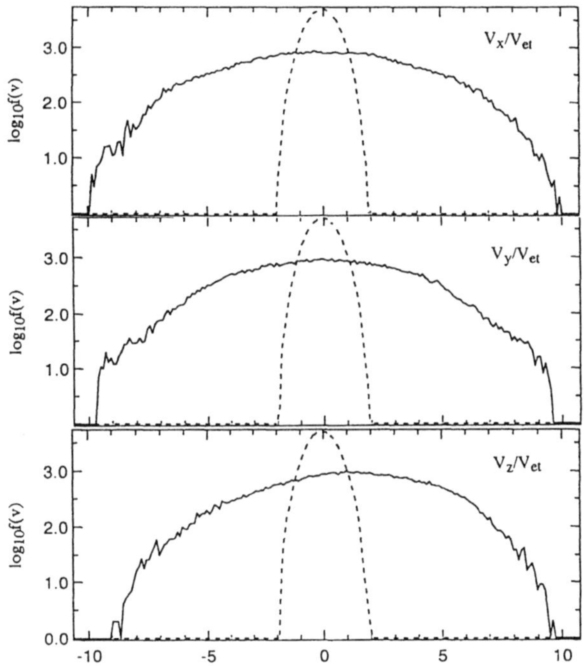

Figure 112. Particle simulation of two-loops coalescence. As Figure 110, but for $\omega_{pe}t = 75$.

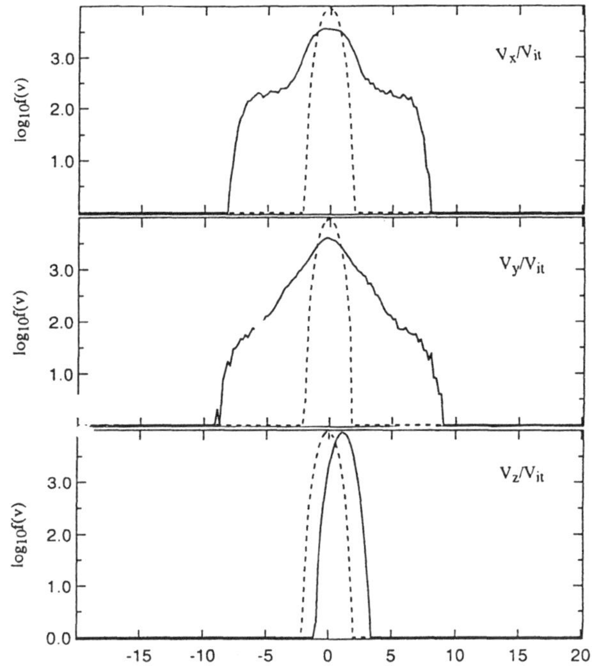

Figure 113. Particle simulation of two-loops coalescence. As Figure 110, but in this case the diagram gives the velocity distribution of *ions* in the interacting region, at $\omega_{pe}t = 20$.

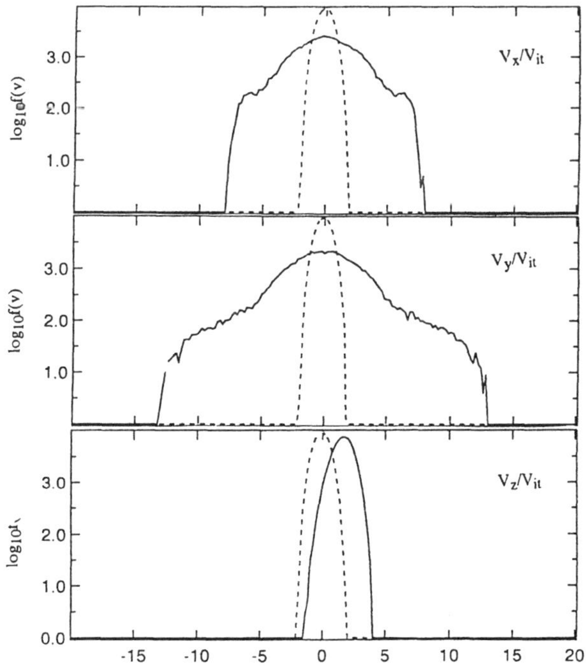

Figure 114. Particle simulation of two-loops coalescence. As Figure 113, but for $\omega_{pe}t = 30$.

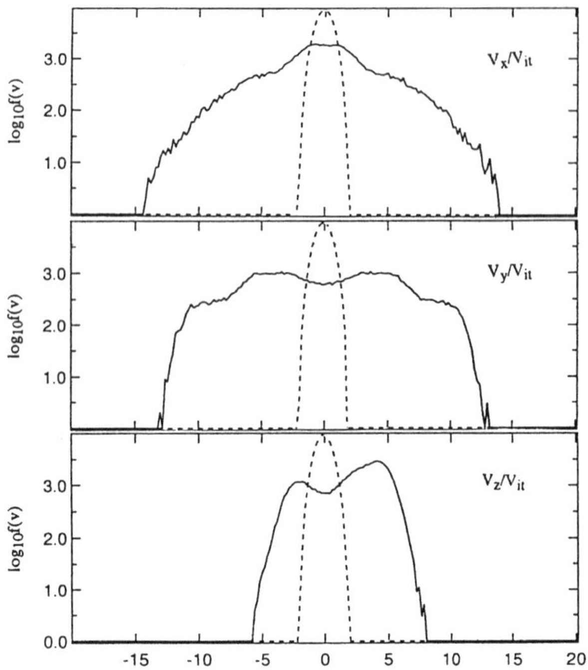

Figure 115. Particle simulation of two-loops coalescence. As Figure 113, but for $\omega_{pe}t = 75$.

current-carrying electrons are decelerated and an electron-current can originate in the region in between the two loops.

Figures 116 to 117 deal with the time history of the electrons that were carrying the original current. It appears that these are accelerated in a direction perpendicular to the magnetic field. Figure 118 shows that a certain fraction of the electrons can be further accelerated in the z-direction. Figure 119 presents the energy spectrum of the electrons that were initially in the region of interaction. After the coalescence, the electrons are additionally accelerated to relativistic energies, through the process of magnetic reconnection.

The simulations described above, show that, during the process of two-loops coalescence with partial magnetic reconnection, electrons are accelerated to relativistic energies. It should still be investigated if rapid acceleration of ions also occurs during two-loops coalescence with complete magnetic reconnection.

5.4. Electromagnetic wave emission from a coalescence region

An important topic in solar physics is to understand the mechanism of electromagnetic wave emission, since many phenomena in solar active regions can be observed primarily at radio frequencies. The type II and III bursts were one of the earliest discovered forms of discrete solar radio emissions (Wild and McCready, 1950). Later, more types of solar microwave emissions from flaring regions have

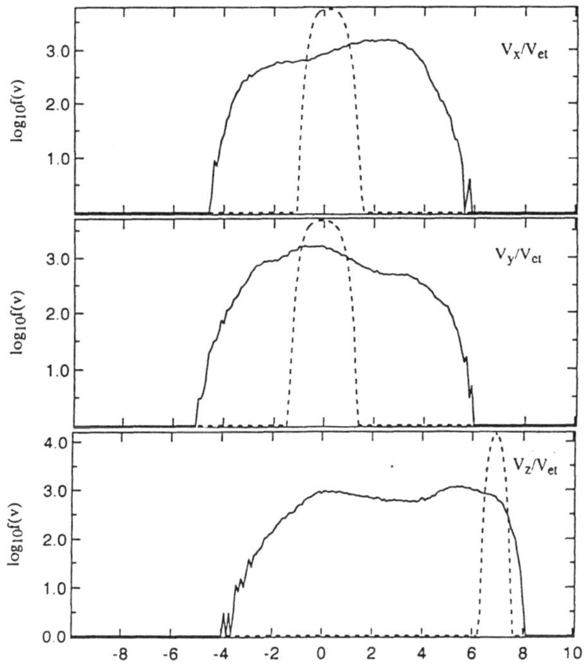

Figure 116. Particle simulation of two-loops coalescence. Velocity distribution of the electrons in one loop at $\omega_{pe}t = 21$. (Zhao *et al.*, 1995b.)

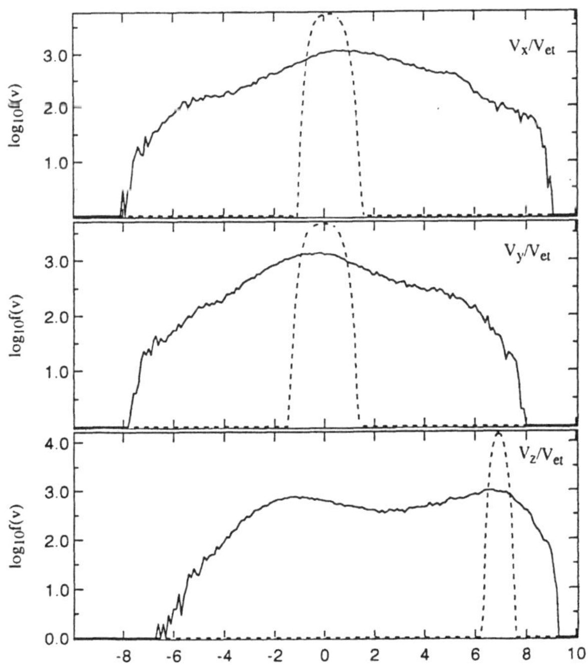

Figure 117. Particle simulation of two-loops coalescence. As Figure 116, but for $\omega_{pe}t = 30$.

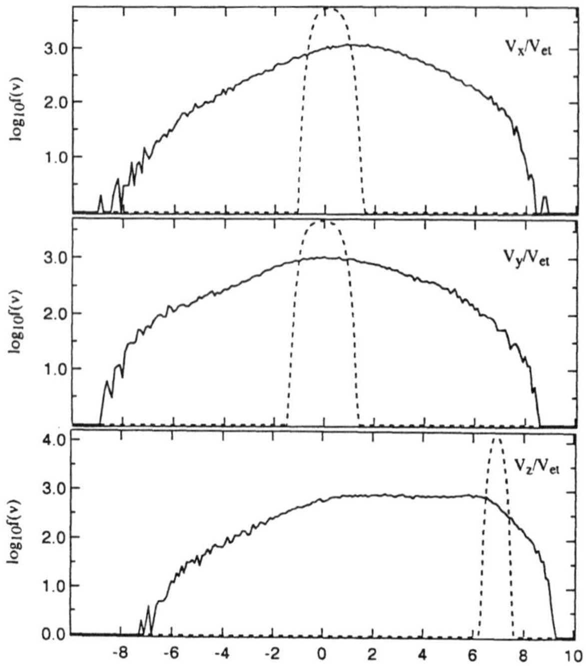

Figure 118. Particle simulation of two-loops coalescence. As Figure 116, but for $\omega_{pe}t = 54$.

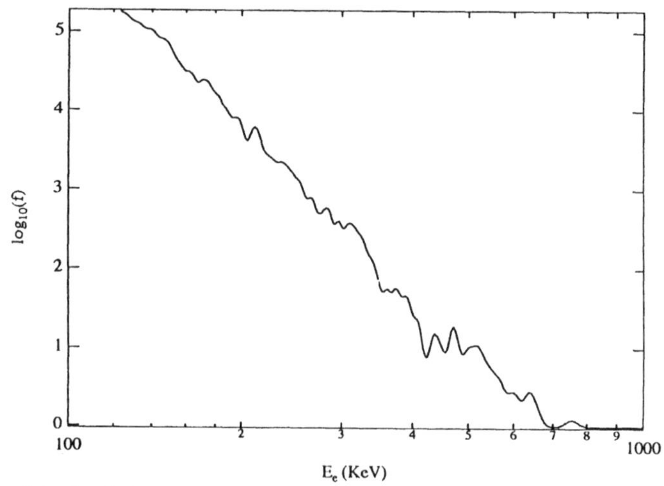

Figure 119. Particle simulation of two-loops coalescence. Energy spectrum of the electrons that were initially in the region of interaction. (Zhao *et al.*, 1995b.)

been identified. Such observations also yield information on the relation between certain emission sources and transient brightenings as seen in soft X-ray images. When X-ray and microwave sources are co-spatial, the microwave source is generally a compact region located near the center of the transient regions seen in

X-ray (Webb *et al.*, 1987). The measurement of a brightness temperature spectrum is important for determining the radiation mechanism responsible for the emission, and for inferring the physical conditions in the source region.

The observation of microwave emissions has provoked great interest in the study of relevant radiation mechanisms. Early investigations studied theoretically and numerically the mechanisms of emission via the coherent cyclotron maser effect in the presence of a loss cone due to a converging magnetic field. These studies suggested the cyclotron maser instability as a source mechanism (Holman *et al.*, 1980; Wu and Lee, 1979; Wagner *et al.*, 1983). This mechanism is able to provide a fair explanation for many of the observed features of auroral kilometric radiation, including the frequency-cutoff of the excited radiation, the polarization and brightness temperature. It explains the continuous spectrum of the emission. The observed burst emission, however, is difficult to be reconciled with the build-up time of a suitable velocity distribution. Therefore, the exploration of other processes for explaining the burst emission is very desirable. Such studies should include the explanation of burst emissions by means of a temperature anisotropy beam instability (Wong and Goldstein, 1990; Winglee *et al.*, 1992), and the direct radiation by runaway electrons in a constant electric field (Tajima *et al.*, 1990).

As was demonstrated in section 5.2, the current-loops can be heated preferentially in the direction perpendicular to the magnetic field, either due to the magnetic pinching effect as well as during current-loop coalescence. This leads to the formation of a plasma with an anisotropic temperature distribution, with $T_\perp > T_{parallel}$. Zhao *et al.* (1995a) used a 3-D fully electromagnetic and relativistic particle code (Buneman, 1993) to study the electromagnetic emission in a plasma with an anisotropy in the electron temperature. The system size used for the simulations is $L_x - L_y = 85\Delta$, $L_z = 160\Delta$, with a million electron-ion pairs, initially homogeneously filling the system. As before, L_x, L_y, and L_z are the lengths of the system along the three Cartesian coordinates and $\Delta (= 1)$ is the grid size. Periodic boundary conditions are used for particles and fields in the z-direction, while radiating boundary conditions (Lindman 1975) are used in the x- and y-directions. The study concentrated on the problem of the character of the electron temperature anisotropy in a column with a radius $r = 8\Delta$, oriented along the z-direction, the axis of which is located in the center of the domain ($x_{cn} = y_{cn} = 43$). The plasma density inside the column is the same as that outside it and the column is taken to be charge-neutral. An electron-temperature anisotropy is introduced with $T_\perp = 16T_{||}$, where $T_\perp$ is the column electron temperature in the direction perpendicular to the column axis and $T_{||}$ is the electron temperature in the direction along the column axis, which is the same as the temperature of the background electrons. Other parameters are as follows: $m_i/m_e = 64$, $T_g/T_i = 1$ (T_g is the background electron thermal temperature and T_i is the ion thermal temperature), $\omega_{pe}t = 0.1$, $\beta = 0.11$, $\lambda_{\text{De}} = 0.469\Delta$, $c/\omega_{pe} = 5.0$, $c/v_{tg} = 10.67$ (v_{tg} is the background electron thermal velocity), $\omega_e/\omega_{pe} = 0.4$. The gyroradii are $\rho_e = 1.17\Delta$, $\rho_i = 9.37\Delta$.

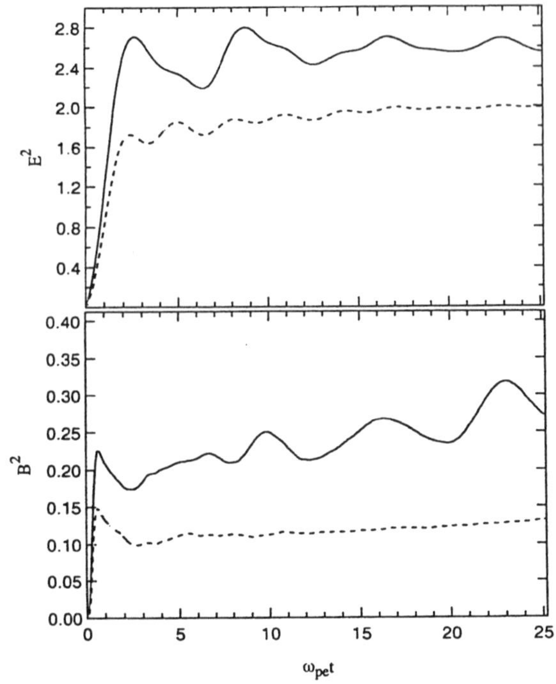

Figure 120. Particle simulation of a current-loop plasma with an imposed temperature anisotropy. The figure presents the time history of the electric and magnetic field energies in the column. (Zhao *et al.*, 1995b.)

Figure 120 presents the time history of the values of the electric and magnetic field energies in the loop. It is clear that the electric and magnetic energies first increase strongly, and thereupon start oscillating smoothly. This electromagnetic instability is driven by an electron temperature anisotropy (Mikhailovski, 1975). Figure 121 plots the time history of the electron kinetic energies $(E_{\|}, E_{\perp})$ in the column. The electrons are initially slightly cooled down in the perpendicular direction, while the electrons are heated up in the parallel direction. The spectrum of B_x in the (ω, k) plane outside the column is illustrated in Figure 122. This latter result was obtained by performing a two-dimensional Fourier transform (one space and one time coordinate). The gray scale in Figure 122 refers to the perturbed intensity. It is logarithmic with an arbitrary zero-point. Apparently, both low-frequency and high-frequency electromagnetic wave are excited: these are propagating nearly symmetrically along the initial magnetic field in both directions. Strong low-frequency waves with frequencies near Ω_e $(0.4\,\omega_{pe})$ are characterized by a right-hand whistler mode as shown in Figure 123, which plots the time evolution of B_x, B_y in the center of the column. This result agrees with the one showing that a whistler mode is excited in a plasma with an electron temperature anisotropy if the plasma frequency exceeds the cyclotron frequency (Mikhailovski, 1975).

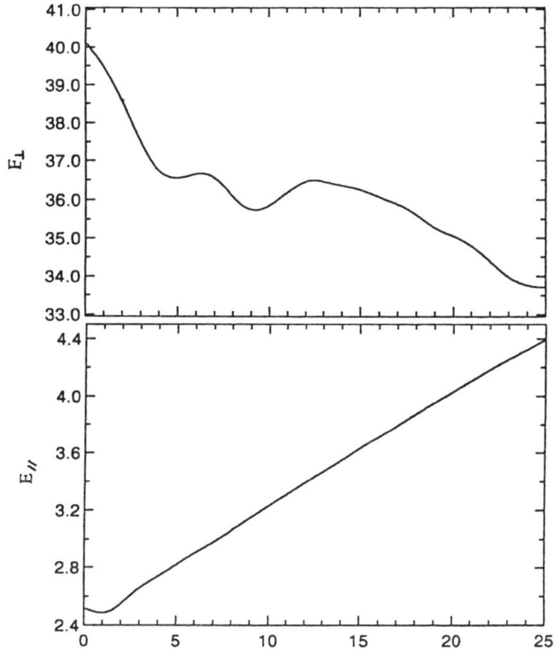

Figure 121. Particle simulation of a current-loop plasma with an imposed temperature anisotropy. The figure gives the time history of the kinetic energy of the electrons in the column. (Zhao *et al.*, 1995b.)

Additional to the low-frequency waves, high-frequency electromagnetic waves are also radiated, with frequencies that are several times the plasma frequency.

In order to obtain an instability due to the temperature anisotropy it is necessary (Mikhailovski, 1975) to have

$$\frac{T_\perp}{T_\|} > 1 + (ck_z/\omega_{pe})^2.$$

A simulation was made with $T_\perp/T_\| = 16$ and $ck_z/\omega_{pe} = 1.5$, values that correspond to the maximum intensity of the excited waves. These values satisfy Mikhailovski's above-given condition, hence the instability does occur. The growth rate of a perturbation (Mikhailovski, 1975) with frequency at the whistler branch being $\omega = \pm\Omega_e/(1 + (\omega_{pe}/ck_z)^2)$ is

$$\gamma = \frac{2\pi^2}{|k_z|}\frac{q}{(1+q)^3}\frac{\omega_B}{(ck)^2}\frac{4\pi e^2}{m}\int\frac{v_\perp^3}{2}\mathrm{d}v_\perp\left[\frac{\partial f}{\partial\epsilon_\perp} + \frac{k_z}{\omega}\left(\frac{\partial f}{\partial v_z} - v_z\frac{\partial f}{\partial\epsilon_\perp}\right)\right]_{v_z=v_{res}},$$

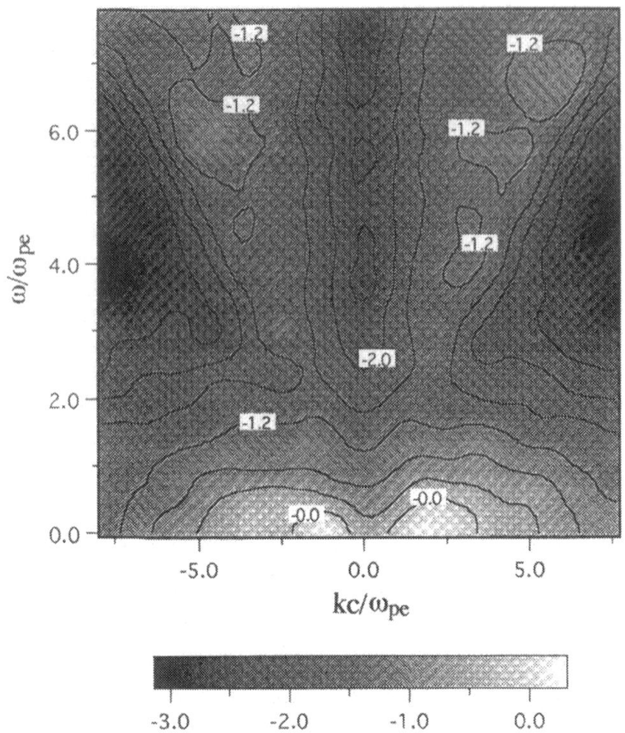

Figure 122. Particle simulation of a current-loop plasma with an imposed temperature anisotropy. The diagram gives the intensity of B_x as a function of (ω, k_z) along the z-direction at $x = 43$, $y = 20$. (Zhao *et al.*, 1995b.)

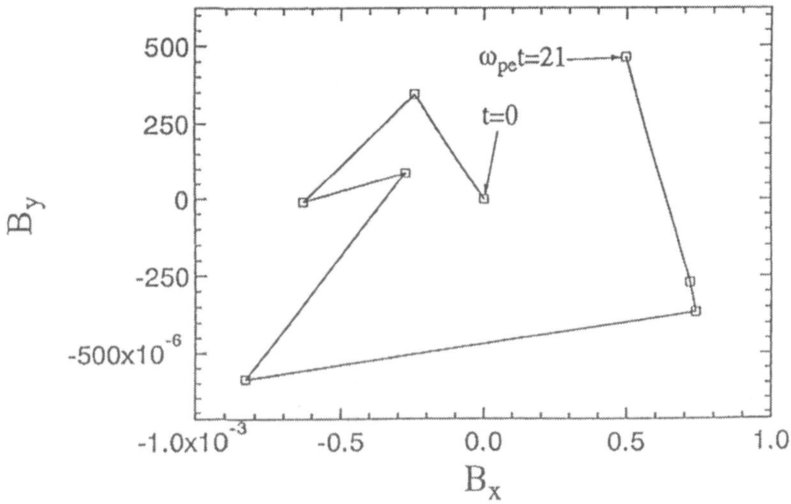

Figure 123. Particle simulation of a current-loop plasma with an imposed temperature anisotropy. The figure gives the time evolution of the (B_x, B_y)-variation in the center of the column. (Zhao *et al.*, 1995b.)

where $q = (\omega_{pe}/ck_z)^2$ and $v_{res} = \Omega_e/k_z(1 + (ck_z/\omega_{pe})^2)$. For a bi-Maxwellian distribution function, which is used as the distribution function in the simulation, with

$$f = n_0 \left(\frac{m}{2\pi T_{\|}}\right)^{1/2} \frac{m}{2\pi T_{\perp}} \exp\left(-\frac{mv_z^2}{2T_{\|}} - \frac{mv_{\perp}^2}{2T_{\perp}}\right),$$

the growth rate becomes

$$\gamma = \frac{\sqrt{\pi}\omega_B^2}{|k_z|v_{T\|}} \frac{q^2}{(1+q)^3} \left[q\left(\frac{T_{\perp}}{T_{\|}} - 1\right) - 1\right] \exp\left(-\frac{\omega_B^2}{k_z^2(1+\frac{1}{q})^2 v_{T\|}^2}\right).$$

It appears from the above equation that the growth time is of the order of $3(\omega_{pe})^{-1}$, which is consistent with the simulation results shown in Figure 120.

The total energy density of the high frequency emission is determined as follows. We define a dimensionless high-frequency energy density, by means of a formula used by Goldman (1984), to define the Langmuir energy-density, as

$$W = \frac{\langle |E|^2 \rangle}{4\pi n_e k_B T_e},$$

where n_e is the electron density and $k_B T_e$ is the electron thermal energy. In the simulation the wave energy of the high-frequency emission appears to be about one tenth of the total wave energy as estimated from Figure 120. An approximate calculation with the simulation parameters given earlier shows that the peak energy density W is about 3×10^{-3}. It is worth remarking that the high-frequency emission driven by the electron temperature anisotropy is as strong as a type III burst emission whose peak energy density exceeds 10^{-5}, expressed in units of the thermal background energy nkT (Goldman, 1984).

Data analyses (Zhao *et al.*, 1995b) of the simulation results presented in section 5.3 gave information on the electromagnetic emission from the interaction region between the two loops just before loop coalescence. The electromagnetic emission process may be a good observational signature for the comparison of observations with results from the simulated process of loop-coalescence.

Some flares that show evidence of two-current loops coalescence with radio emission have been observed simultanously by the soft X-ray telescope of *Yohkoh* and the Nobeyama radioheliograph. One of the examples (Hanaoka, 1994) is the flare of 15 July 1992, 03:55 UT (cf., also section 6.3). In the soft X-ray images (cf., Figure 129) one observes in the pre-flare stage a bird-shaped diffuse bright image with wings widely spread out. The two wings are the two loops and the body is at the position where the two loops interact. As time goes on, the point of interaction strongly brightens up. After having reached maximum intensity the brightness of the body starts to decrease while simultaneously the brightness of the wings gradually increases. This phenomenon looks as if the two loops collided with each other and heated the region of interacting. In turn, the interaction energy seems to have heated the loops. The radio images show a shape reminiscent of two-loops interaction. It seems that the primary radio source is associated with the region of interaction.

Lim *et al.* (1993), investigating the microwave burst emission from flares, also indicated that some radio emission comes from the interacting region of loops. These observations suggest that certain radio emissions from flares may be radiated by colliding loops.

5.5. Collision between a Plasma Cloud and a Current-Loop

The formation of plasmoids during the process of coalescence has been widely investigated in theories and simulations (Sakai, 1989; Sakai, 1990; Forbes and Malherbe, 1991; Fushiki and Sakai 1995). Such studies demonstrate that two plasmoids, generated at both sides of an X-point, may escape from the region of interaction. Observations of the flare of 2 December, 1991 (cf. section 6.7.2 and Figures 149 and 150) present an impressive picture of a plasmoid being ejected from the region of the X-type neutral sheet, in association with the process of magnetic reconnection (Tsuneta, 1993). In order to interpret the existence of an impulsive hard X-ray source at the loop-top in solar flares, Sakai *et al* (1995) simulated a collision of a plasmoid with a force-free current-loop by using a 3-D MHD code. The results show that another current-loop is induced in front of the plasmoid while it is moving towards the force-free current. This secondary current-loop coalesces with the force-free current associated with the process of magnetic reconnection. They concluded that a collision between a plasmoid and a current may trigger a flare.

Here we describe a study of the same phenomenon with a 3-D electromagnetic particle code (Buneman, 1994), undertaken in order to investigate plasma heating as well as particle acceleration in such a collision process (Zhao *et al.*, 1995c). The simulation model is a simplified version of a collision between a plasma cloud and a current-loop.

The system domain is a mesh of size $130\Delta \times 100\Delta \times 140\Delta$, where Δ $(=1)$ is the grid size. A periodic boundary condition is used for particles and fields in the z-direction, while a radiating boundary condition (Lindman, 1975) is used in the x- and y-directions. Particles that hit these boundaries are stopped there and 50% of them are reflected randomly with a thermal velocity. There are approximately 4 million particles (electrons and ions). They uniformly fill the entire domain and keep the domain charge-neutral.

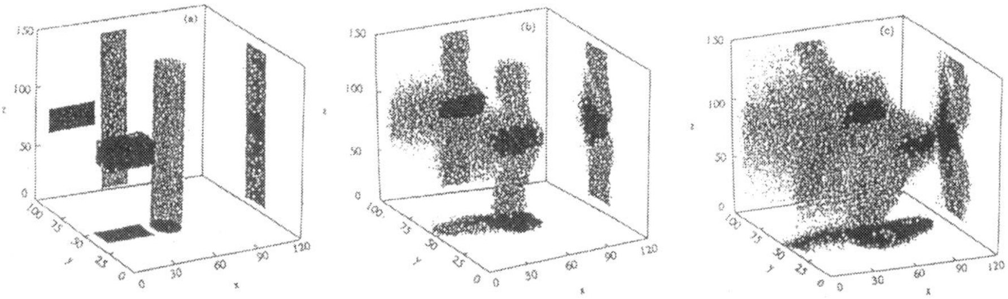

Figure 124. Time history of the collision between a plasma cloud and a current-loop. Current-carrying electrons and electrons in the plasma cloud are plotted in 3-D space, along with their projections onto the xy-, xz-, and yz-planes, at times $\omega_{pe}t =$ (*left*) 0, (*middle*) 30, and (*right*) 60. (Zhao *et al.*, 1995c. From *Astrophys. J.*)

An electron-carrying current-loop oriented in the z-direction is specified in a column with a radius $r = 10\Delta$, with the center located at $x_{cn} = 60$, $y_{cn} = 50$ in the xy-plane. The plasma density within the column is equal to the density outside it. The electron drift velocity of the loop (v_d) is the same as the background electron thermal velocity (v_{et}). Next to the current-loop a cloud of electrons and ions is introduced; it has a size of $35\Delta \times 16\Delta \times 20\Delta$, and its center is at the position $x_{cn} = 28$, $y_{cn} = 50$, $z_{cn} = 70$. Hence, the front of the cloud is at a distance of 5Δ from the loop. The density of the cloud relates to the background plasma density as $n_c/n_g = 8$. The cloud is pushed across a homogeneous ambient magnetic field B_z into the x-direction with a bulk velocity of $5v_A$, where v_A is the Alfvén velocity. Other parameters are: $m_i/m_e = 64$, $\Omega_c/\omega_{pe} = 0.4$, $c/v_{et} = 10.67$, $\beta = 0.111$, $\lambda_{\mathrm{De}} = 0.469\Delta$, $c/\omega_{pe} = 5.0\Delta$. The Alfvén velocity in the simulation is $v_A/c = (m_e/m_i)^{1/2}\Omega_c/\omega_{pe} = 0.05$. The gyroradii of the electrons and ions are $\rho_e = 1.17\Delta$, and $\rho_i = 9.36\Delta$, respectively.

The simulation ran up to a simulated time of $60\omega_{pe}t$, which corresponds to approximately 4 electron gyroperiods. This guarantees that the electrons are thoroughly magnetized. However, this period is equivalent to 0.06 ion gyroperiods, hence the ion-gas is still unmagnetized.

Figure 124 shows the time history of the current-carrying electrons and the electrons in the cloud in 3-D space with their projections onto the xy-, xz-, and yz-planes. The background particles are not plotted in the Figure. From the given initial conditions, it is easy to calculate that the cloud will collide with the loop at $\omega_{pe}t \approx 4$. After the collision the cloud penetrates into the loop and it continues to move in the same direction. The expansion of the electron sheath at the rear end of the cloud along the ambient magnetic field is particularly noticeable, and has thoroughly been investigated by Neubert *et al.* (1992). The evolution of the collision process is also clearly seen in the time history of B_z to B_x, which are projected onto the plane at $y = 50$, and plotted as arrows in Figure 125.

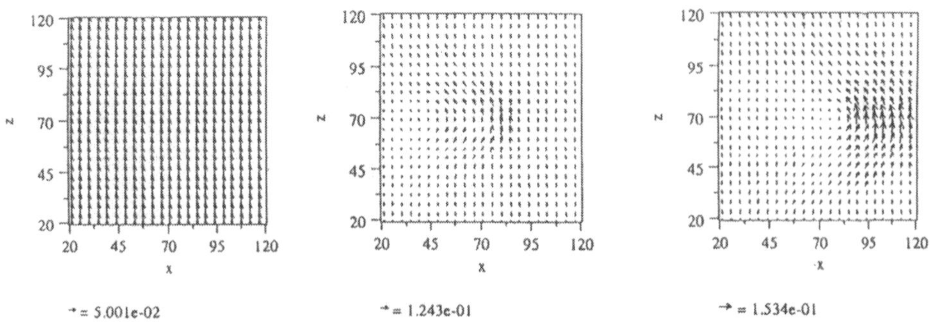

Figure 125. Collision between a plasma cloud and a current-loop. The figure presents the time history of the B_z- to B_x- components onto the plane $y = 50$ by plotting them as arrows, at the times $\omega_{pe}t =$ (*left*) 0, (*middle*) 30, and (*right*) 60. (Zhao *et al.*, 1995c. From *Astrophys. J.*)

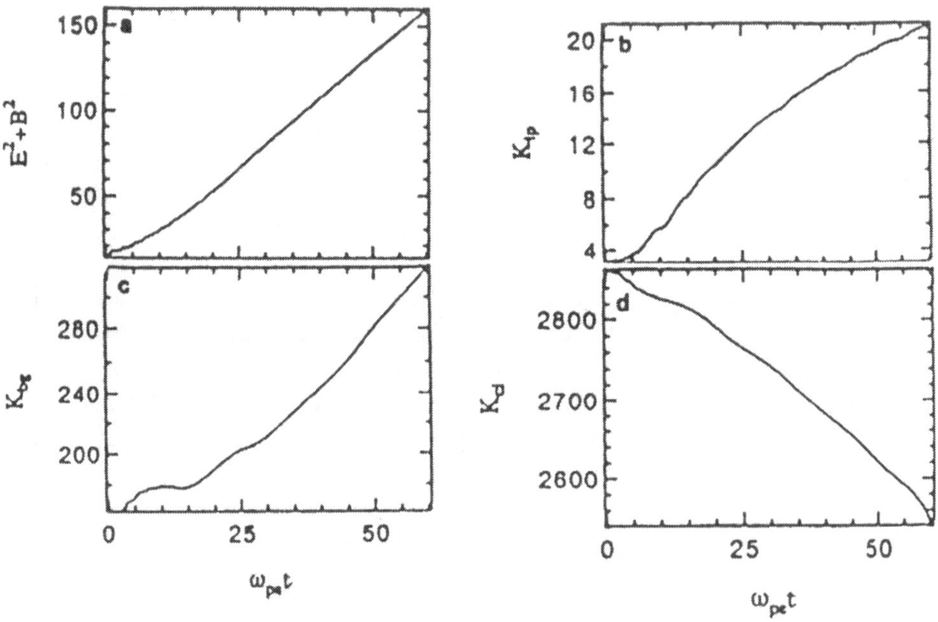

Figure 126. Collision between a plasma cloud and a current-loop. Time history of the field energy and the kinetic energy components. (a) Field energy; (b) loop kinetic energy; (c) background kinetic energy; (d) cloud kinetic energy. (Zhao *et al.*, 1995c. From *Astrophys. J.*)

The ambient magnetic field appears to be strongly compressed and enhanced in the head of the cloud. The enhanced region shifts, together with the movement of the cloud.

Figure 126 plots the calculated time histories of the field-energy and the kinetic energies of the particles in the loop, cloud, and background. The ambient magnetic field is subtracted in calculating the field energy. It should be noted that saturation has apparently not yet been reached, although that is expected to occur when the magnetic energy density equals the cloud pressure. In order to achieve saturation,

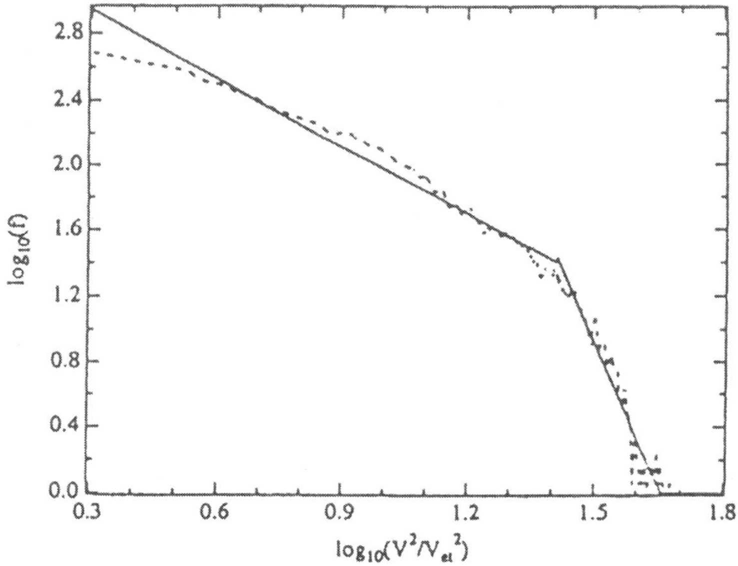

Figure 127. Collision between a plasma cloud and a current-loop. Energy spectrum for electrons in the loop, obtained from the simulation described in the text (*dashed line*), and its linear fitting (*solid lines*). The diagram is for the time $\omega_{pe}t = 60$. The energy is normalized by the thermal energy. (Zhao *et al.*, 1995c. From *Astrophys. J.*)

a longer simulation time is apparently needed, but that would go beyond the acceptable limit set by the computer used.

Figure 126 shows that the kinetic energy of the cloud decreases, opposite to a simultaneous increase of the field energy and of the kinetic energies of loop and background. Since the cloud is impeded by the loop, the energy lost is converted. It deforms the loop and compresses the ambient magnetic field. Hence their energies increase. This simulation result demonstrates that the amount of released energy is in good agreement with the gains of the energies of the field, loop, and background, respectively. At the end of the simulation, the kinetic energy of the loop has increased by a factor 7 as compared with its initial energy. Even though we see that the kinetic energy of the cloud decreases, it is, strictly speaking, a release of the energy lost in the drift direction, coupled with heating of the cloud in other directions.

A calculated energy spectrum of the electrons in the loop is displayed in Figure 127 for the time $\omega_{pe}t = 60$. The energy (horizontal axis) is on a logarithmic scale and normalized by the thermal energy. The dashed line shows the result obtained from the simulation. The spectral index along this curve varies from 1.38 to 5.87. A linear fit by a broken power law has been tried; it appears that the simulation can be represented by the two solid lines drawn in the figure. We therefore conclude that the energy spectrum of the non-thermal electrons can be well represented by

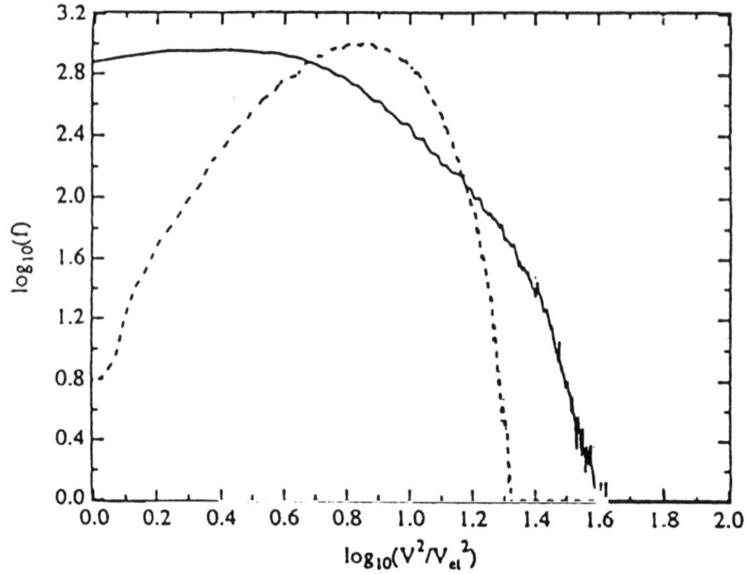

Figure 128. Collision between a plasma cloud and a current-loop. Energy spectrum of electrons in the plasma cloud at $\omega_{pe}t = 0$ (dashed line) and $\omega_{pe}t = 60$ (solid line). (Zhao *et al.*, 1995c. From *Astrophys. J.*)

two power-law spectra. A further investigation shows that the bending of the loop is apparently responsible for the broken power law.

These calculations are supported by the observation that broken-power law spectra are a common feature of hard X-ray flares (Tanaka *et al.*, 1984; Winglee *et al.*, 1991; Inda-Koide, 1994).

In order to show the time development of the energy of the electrons, Figure 128 plots the energy spectrum of the electrons in the cloud at two times, viz., at $\omega_{pe}t = 0$ (the dashed line) and $\omega_{pe}t = 60$ (the solid line). As compared with the initial situation the spectrum becomes broader with time due to thermal heating. The maximum energy is roughly 30 times the thermal energy, as shown both in Figures 127 and 128. The ratio levels off with time, since the spectra do not alter much after $\omega_{pe}t = 40$, although saturation had not been reached at that time.

The two populations of electrons that are appearing in Figure 127 correspond well with the soft and hard X-ray sources in the loop observed in the flare of 13 January, 1992 (Masuda *et al.*, 1994). On the other hand, even though the electrons in the cloud do not display a broken spectrum in Figure 128, they should presumably correspond to the hard X-ray sources above the loop apex, also described by Masuda *et al.* (1994). Since the high-energy electrons in the loop and cloud appear to be almost identical, having the same maximum energies, we conclude that the combined electron populations in the simulation are capable of emitting both soft- and hard X-rays, with observed energies that agree approximately with the results of the simulation (cf., further also section 6.7.3).

6. Observations of Flares in Relation to Loop Interactions and Loop Instability

6.1. INTRODUCTION

In the review paper by Sakai and Ohsawa (1987) observations of two flares (those of 7 June, 1980 and of 26 November, 1982) were compared with results of predictions of current-loop coalescence (Tajima and Sakai, 1989a, b) and with results from 2-D simulations (Tajima *et al.*, 1987). The current-loop coalescence process provides keys for understanding several characteristics of solar flares, such as their impulsive nature, simultaneous plasma heating, high-energy particle acceleration, amplitude oscillations of electromagnetic emission, as well as the characteristic development of microwave images obtained during flares (this applies notably to the flare of 26 November, 1982). The observed double peaks in the quasi-periodic oscillations in the time-profiles of the above two flares could well be explained in terms of coalescence of two current-loops.

We had, however, to wait for improved high-resolution telescopes, yielding sharper soft- and hard-X ray images, to find clear and unambiguous evidences of two-loops interactions. *Yohkoh* gave new impetus to the observational study of such interactions. Several evidences (Takahashi *et al.*, 1995; Inda-Koide *et al.*, 1995; de Jager *et al.*, 1995; Roumeliotis, 1995) showing two-loops coalescence in small flares, were observed by *Yohkoh*. Many other loop-interaction processes in coronal active regions were found in association with transient X-ray brightenings (Shimizu *et al.*, 1992, 1994). Other observations refer to single-loop instabilities, and to cusp-type flares. In the subsections of this Section we discuss some observational evidences of single-loop instability and of two-loops interactions.

We have arranged these subsections in the context of the theoretical developments and flare classification scheme presented in the previous Sections. Accordingly, we proceed by first describing flares based on Y-, X-, and I-type interaction. Such flares can either be confined or eruptive, because such interactions are not limited to one of the two main types. Thereafter, we describe single-loop flares and then the cusp-type flares. We discuss them in that order because we consider the single-loop flares as a basic type of the confined flares (even when some of them may appear to be unresolved two-loops flares), while we consider the cusp-type flares as an interesting basic form of the eruptive/dynamic flares. Finally, we describe a fully developed eruptive/dynamic flare.

6.2. GENERALITIES ON TWO-LOOPS INTERACTIONS IN ACTIVE REGIONS AND POST-FLARE LOOP SYSTEMS

Shimizu *et al.* (1994) investigated 142 transient brightenings in active regions observed by *Yohkoh*'s soft X-ray telescope during a period in October 1991. They classified them in terms of morphology and time evolution as follows: (1) simultaneous multiple-loop brightenings (57 events) are seen as frequently as brightenings of single (59 events) and more frequently than point-like (26 events) structures; (2) for multiple-loop brightenings, the loops tend to brighten starting from their footpoints and/or from the apparent contact point in the initial phase of transient brightenings, followed by a brightening of the entire loops; (3) more than one-half of the multiple-loop brightenings are Y-type configurations, in which the contact points are located close to their footpoints.

Shimizu *et al.* (1994) also showed that *two* brightening loops are seen in 86% of the multiple-loop events and 14% of them show *more than two* loops. They also found that the multiple-loop interaction processes can be classified into four types; Y-type (61.4%), X-type (24.6%), I-type (1.7%), and others (12.3%). Another conclusion is that some of the apparent single-loop and point-like brightenings may be multiple-loop structures on sub-telescopic scales and that most of the transient brightenings may consist of multiple loops.

Smartt *et al.* (1993) and Airapetian *et al.* (1995) investigated the occasional interactions of post-flare coronal loop systems by observing high-resolution images in visible coronal lines, i.e., Fe XVI (5303 Å) and Fe X (6374 Å). The observations reveal transient enhancements that characterize localized loop-interactions. The interactions evidently result in a localized energy release that causes the observed enhancements; a result based on ninety observed events. They concluded that the phenomenon of post-flare loop-interaction can be interpreted as a process of localized loop-coalescence involving partial magnetic reconnection. As the plasma β in the corona is about unity for the above events, the coalescence process is not explosive, as predicted by Sakai and de Jager (1989b, 1991). This observation is an evidence of coronal loop-interaction in the post-flare phase observed in visible coronal lines. At about the same time, Šimberová *et al.* (1993) presented evidence of large-scale two-loops interactions in the corona from *Skylab* soft X-ray images (the limb event of 13–14 August, 1973). They concluded that the giant arches that are occasionally observed above eruptive flares may originate through interactive reconnection of large-scale magnetic fields which form loops high in the corona. These loops are approaching each other, following changes in the coronal structure caused by lower situated eruptive flares. The schematic drawings of the loop structure and interaction, given in Figure 6 of the quoted paper, imply that complete magnetic reconnection occurred during two-loops coalescence after the appearance of the newly formed loops.

6.3. Two-Loops Interaction: A Case of an Y-Type Flare

The flare of 15 July, 1992 (Hanaoka, 1994; Nakajima, 1994) seems to be a case of Y-type interaction but not in its 'pure' form (Figure 129). The flare occurred at 03:55 UT, but two loops were already visible one day earlier. They had their western footpoints near a sunspot (Figure 129(a)). At that time a protrusion of the spot moved westward ($v = 0.25\ \mathrm{km\,s^{-1}}$) while one of the loops was dragged

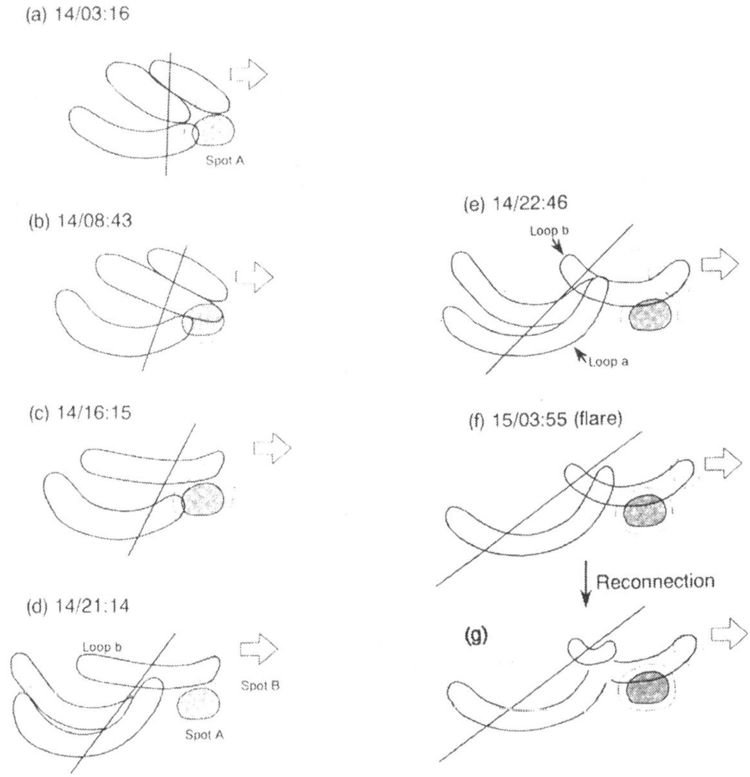

Figure 129. The flare of 15 July 1992. Schematic drawings of the relation between the coronal loops, the spots, and the magnetic neutral line (straight line). The arrows denotes directions of photospheric motions. (Hanaoka, 1994. From *Astrophys. J.*)

along. Thus the shear increased, which also resulted in a gradual change of the orientation of the inversion line (solid line in the figures).

The flare occurred at the intersection of the sheared loops (a) and (b). The total thermal energy content of the flare showed a rapid increase during the first 3 minutes, apparently during the impulsive phase. It seems that the flare was due to complete reconnection between loops (a) and (b), because there were traces of a small loop after the flare (Figure 129(g)). That loop seems to have originated during the flare. The question whether this flare fully satisfies the theoretical predictions for Y-type reconnection has not yet been discussed.

Another case of a Y-type flare is the one of 11 August, 1992, described by Farnik, Hudson and Watanabe (1996); cf., section 2.2.2. and Figure 8 of this review.

6.4. Two-loops Interaction; a Case of an X-type Flare

The limb flare observed by *Yohkoh* on 2 August, 1993 at 08:30 UT (GOES classification B6.8) is a good example of partial X-type interaction (de Jager *et al.*, 1995).

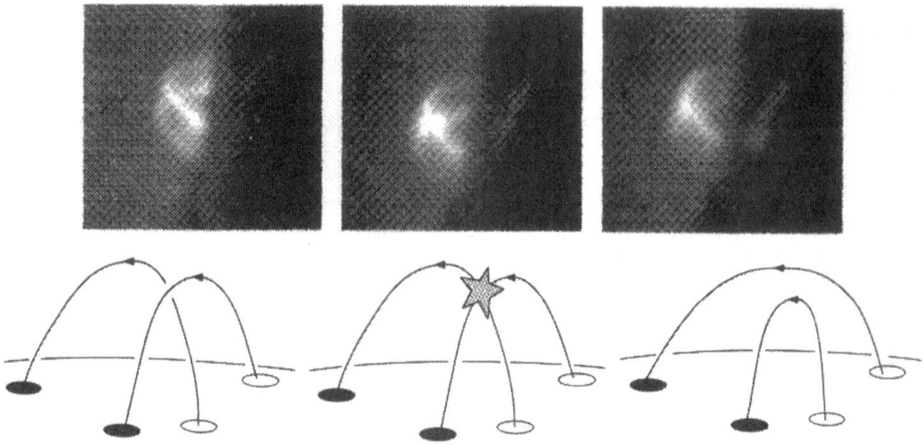

Figure 130. Upper frames: soft X-ray observations of the limb flare of 2 August, 1993 at 08:25:10 UT (*left image*), 08:29:26 (*middle*), and 08:34:46 UT. The flare appears at the crossing point of the two loops. *Yohkoh* SXT observations through a filter centered at $\simeq$ 1.0 nm wavelength, with exposure times 5.3 s (*left*) and 1.3 s (*middle* and *right*). (de Jager *et al.*, 1995. From *Solar Phys.*). *Lower frames*: reconstructed images, showing the loops before, during and after the flare, at three times that are not exactly identical to those of the upper frames. (Rouzeliotis, 1995. From *Astrophys. J.*)

The flare was situated at the Sun's eastern limb in active region 7558. Its general appearance was that of two crossing loops (Figure 130), of which the two southern footpoints had the same magnetic polarity, opposite to that of the northern footpoints. A brightening at the intersection of the two loops occurred simultaneously with brightening at the four footpoints, and was followed by a flare at the loops's intersection.

Preflare activity: between about 08:04 and 08:10 UT there was a general drop in the intensity in all the pre-footpoint areas, as well as in the cross-point area, by some 30 to 40% (Figure 131). More dramatic was the simultaneous intensity decrease of a surge, situated at some distance ($\approx$ 20 000 km) north of the flaring area. Its intensity went down by 80% during a 4-min interval at the same time. Thereafter, the surge intensity continued to decline but at a much reduced rate, until 08:41 UT. At that time the flare ended, and simultaneously the surge rapidly brightened again.

At 08:10 UT the decrease of both footpoint and flare intensity was followed by a sharp rise, back to the earlier level, till 08:20 UT when the four footpoints as well as the flare area started brightening up, initially slowly, later quicker. This is a clear example of flare precursor activity.

Flaring period: a very weak hard X-ray burst was observed by the hard X-ray telescope aboard *Yohkoh* at 08:26.0 UT. It lasted only for a few seconds. Already six minutes earlier, the loops as well as the flare had started to brighten up in soft X-ray images (Figure 130). A sudden sharp soft X-ray brightening of the footpoints started at 08:22.8 UT. At the same time the two loops which were apparently connecting

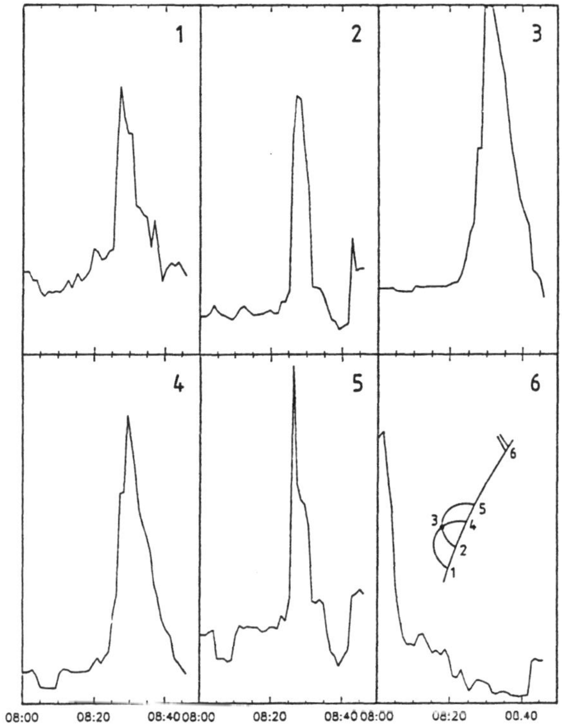

Figure 131. Flare of 2 August, 1993. Normalized soft X-ray countrates as functions of time for six points in the flare area. For allocation of the six numbers, reference is made to the lower right subfigure. The coordinate-values are all in the same units. (de Jager *et al.*, 1995. From *Solar Phys.*)

the footpoints, became visible. Figure 131 shows the normalized countrates as functions of time for six selected areas: those of the four footpoints (numbered 1, 2, 4, 5), the cross-point of the loops (i.e., the flare, No. 3), and the surge (No. 6). The loops were about equally bright near footpoints 1 and 5; those at footpoints 2 and 4 were brighter. The average of the footpoints soft X-ray brightness reached its maximum values at 08:28.1 UT. This brightening had ended at about 08:40 UT.

By an image-deconvolution method Roumeliotis (1995) reconstructed sharper images of the event. Thse are shown in the lower frames of Figure 130. They present three phases: (1) unconnected and (2) reconnected loops during the event and (3) detached loops after it. The latter situation is one of lower energy than the former two, and is therefore a situation to be expected theoretically.

We further note some other characteristics:

– The footpoints and 'flare' started brightening up simultaneously.

– The footpoints, though, reached maximum intensity two minutes earlier than the flare. Hence, the flare intensity curve is somewhat delayed as compared to those of the footpoints.

– The weak hard X-ray burst occurred two minutes earlier than the loop and footpoint maximum, but well after the start of the footpoint and flare brightening.

Temperature determinations show that the maximum temperature of the plasma in the flare knot and loops was about 2.2 times the background value (de Jager *et al.*, 1995). Our Figure 62 shows that in the case of X-type reconnection one may expect only one peak in the temperature during the event (contrary to the cases of Y- and I-type reconnection). A comparison of the above observations, coupled to determinations of electron temperature and emission measure of the footpoints and cross points, with the predictions of reconnection theory as shown in Figure 62, yields that reconnection involved a magnetic field of a few gauss only, which is a few percent of the local photospheric field component. This conclusion shows that we are dealing here with partial reconnection (in which only a part of the B-field reconnects). It supports the expected filamentary (fluxthread) character of the interacting loops (cf., our Figure 31). Another result is that the sequence of events is not compatible with one act of partial reconnection only; it rather seems that there was a continuous series of consecutive small-scale interactions in the course of time. Combining these evidences, the hypotheses was forwarded that a series of fluxthread interactions yielded a steady increase of the current-strength in the loops till the flare originated.

To test this hypothesis, Sakai and de Jager (1996) investigated the above case of X-type loop-coalescence by using a 3-D resistive MHD code (Fushiki and Sakai, 1995b). The left-hand set of diagrams in Figure 132 shows the time-development of the X-type loop coalescence, in which the surfaces of equal density ($\rho = 2.0$) are shown. The right-hand set shows three isosurfaces, that of current-intensity $|\mathbf{J}| = 0.029$, of magnetic field strength $|\mathbf{B}| = 0.707$, and of equal velocity $|\mathbf{v}| = 0.206$, all at $t = 4.5\ \tau_{\mathrm{A}}$. In the lower diagram in the right-hand column we observe the plasma flow caused by the partial magnetic reconnection near the coalescence region. Figure 133 shows the predicted time-history of the temperature and density near the center of the coalescence region. The calculations were made assuming for the plasma $\beta = 0.08$. The consequent temperature enhancement is $\simeq 2.2$ times the background value (cf., the figure), which agrees well with the observations. This temperature enhancement can be explained only by partial magnetic reconnection, not by adiabatic heating.

One of the main conclusions, viz., that of *ongoing partial magnetic reconnection*, therefore confirms a conclusion reached earlier for other flares by Tsuneta (1993, 1994). The important other aspect of this flare, brought forward by Roumeliotis (1995), viz., detachment of the loops after the flaring period, has not yet been simulated in theoretical studies.

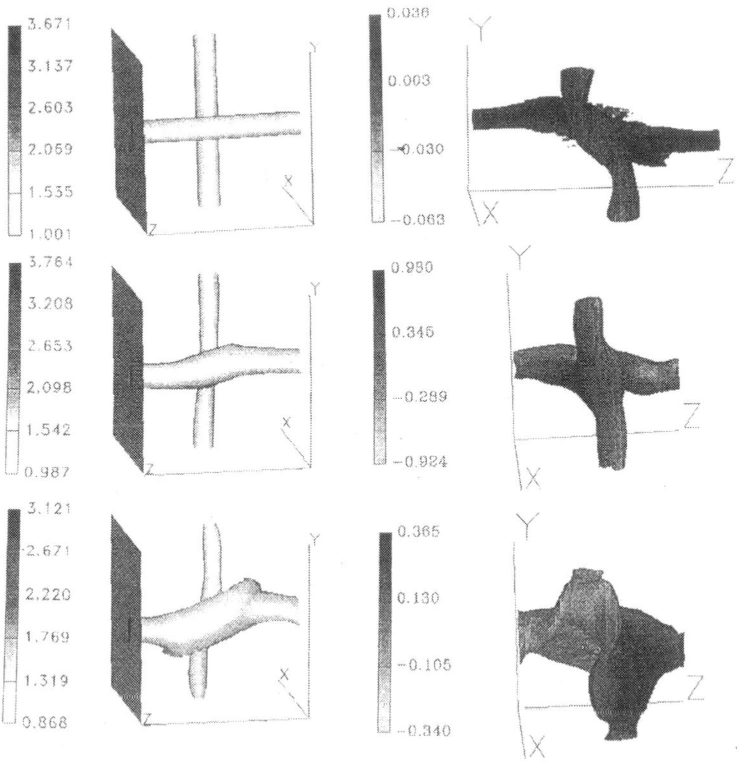

Figure 132. Simulation of the flare of 2 August, 1993. The left-hand set of images shows the time development of the isosurfaces of density for $\rho = 2.0$, at $t = 0$ (*above*), 2 (*middle*), and $4.5\tau_A$ (*below*), respectively. The right-hand images show the isosurfaces at $t = 4.5\tau_A$ of the current-intensity $|\mathbf{J}| = 0.029$ (*above*), the magnetic field strength $|\mathbf{B}| = 0.707$ (*middle*), and the velocity $|\mathbf{v}| = 0.206$ (*below*), respectively. Units are arbitrary. (Sakai and de Jager, 1996.)

6.5. Two-loops Interaction; Cases of I-type Flares

6.5.1. *Introduction*

According to a statistical study by Shimizu *et al.* (1994) I-type reconnection is fairly rare. Yet, the process has interesting theoretical aspects, and therefore we discuss it here in some detail. As reviewed by Sakai and Ohsawa (1987), Tajima *et al.* (1987) studied the process of explosive magnetic reconnection driven by current-loop coalescence and they simulated explosive and quasi-periodic particle acceleration. They applied this model to an explanation of the quasi-periodic (double-peak) structures in some hard X-ray flares (Kane *et al.*, 1983; Forrest and Chupp, 1983; Nakajima *et al.*, 1983; Kiplinger *et al.*, 1983). These flares may have been caused by I-type collisions of loops. In this section we discuss a few more events that seem to be due to I-type current-loop interaction.

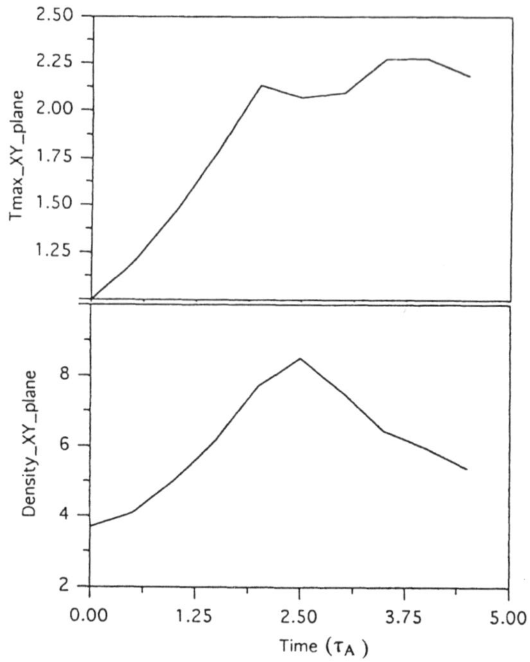

Figure 133. Simulation of the flare of 2 August, 1993. The figure shows the time history of the temperature (*above*) and density (*below*) near the center of the coalescence region. Temperature and density are normalized to the values before coalescence. (Sakai and de Jager, 1996.)

6.5.2. *Type III Radiobursts as I-type Reconnection Indicators*

Aschwanden *et al.* (1994b) found, from the analysis of solar type III bursts observed during a flare on 29 March, 1980, that particle acceleration and injection into magnetic structures occurs intrinsically in a pulsed mode with a typical period of 1 to 2 s. They find evidence that these processes find their origin in a single, spatially coherent, nonlinear system, rather than in a stochastic system with many spatially independent components (upper two frames of Figure 134). They also found that the quasi-periodic acceleration process during loop-coalescence may explain the observed periodicity in solar type III bursts.

Another interesting observation of type III radiobursts, emitted by a flare on the same day as the preceding one, was studied by Aschwanden *et al.* (1994b). The fairly strict periodicity of the bursts is noteworthy (Figure 134, lower frame). These observations seem to confirm the predictions presented in Figures 54 and 60. We therefore classify these flares as possible cases of I-type reconnection is spite of lacking information on the loop structure of this event. The observed series of pulsed type III radio bursts are associated with electrons accelerated to velocities of around $c/4$ to $c/2$, propagating along magnetic field lines. The observations

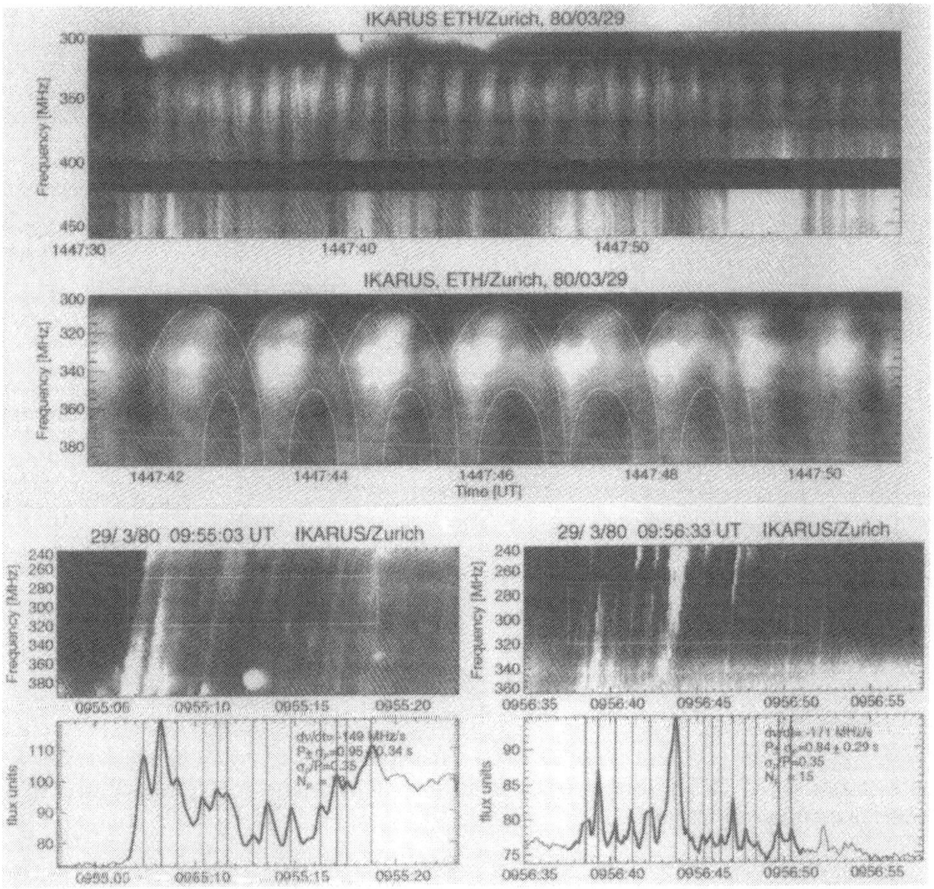

Figure 134. Upper two panels: periodic radio emission, recorded by the digital spectrograph IKARUS of ETH/Zürich on 29 March, 1980, 14:47:30–14:48:00 UT, at 300–450 MHz. A 10 s segment (box in the upper frame) is expanded in the bottom frame. White areas represent enhanced emission. The pulsating structures show type III-like drifting structures, while some of the structures exhibit inverted U-type characteristics (contoured by white lines in the lower frame). *Lower panels*: type III bursts in an earlier flare on 29 March, 1980 at 09:56 UT, showing a fairly strict periodicity. (Aschwanden *et al.* 1994a. From *Astrophys. J.*)

evidently suggest motions along a loop and the repeating signature refers to a repeated series of accelerations. This is fully in agreement with what one would expect for I-type reconnection (see Figure 60). The further analysis by Aschwanden *et al.* (1994b), specifies the characteristics of the site of acceleration, as described above. This result also supports the hypothesis of I-type coalescence. In the quoted paper the authors present five more similar cases.

Karlicky *et al.* (1993) suggested that the 'necklace' structures of narrow-band dm-spikes observed during the flare of 27 April, 1981 were caused by coalescence between two interacting flare loops. The phenomenon has some resemblance to what happens in I-type reconnection.

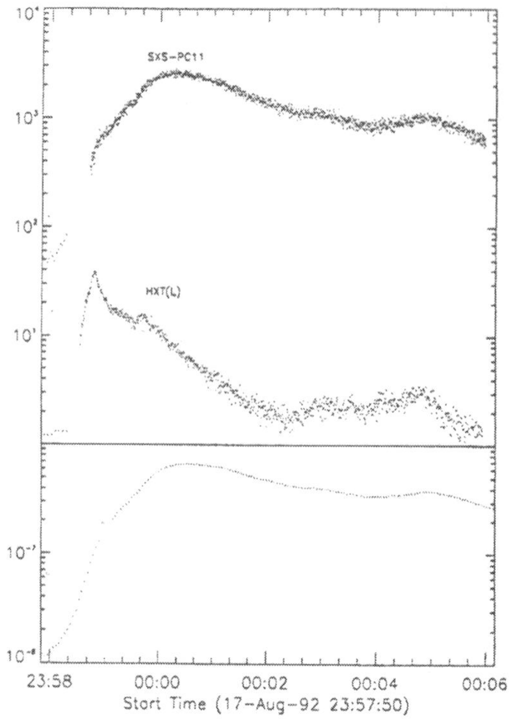

Figure 135. Flare of 17 August, 1992, 23:58 UT. From *above* to *below*: the *Yohkoh* lightcurves in the 2–15 keV and 14–23 keV ranges, and the GOES (0.5–40 Å) lightcurve. (Takahashi *et al.*, 1995. From *Publ. Astron. Soc. Japan.*)

6.5.3. *The Flare of 17 August, 1992, 23:58 UT*

The *Yohkoh* satellite observed the complete development of a flare on 17 August, 1992 at 23:58 UT, from its onset to the decay. It was an impulsive C4.3 flare in the GOES 1–8 Å intensity-scale and it showed a double peak in the soft X-ray energy ranges (Takahashi *et al.*, 1995). This flare was also observed by the magnetograph of the solar flare telescope at the National Astronomical Observatory of Japan at Mitaka (Ichimoto *et al.*, 1991), as well as by the Nobeyama Radioheliograph (Enome *et al.*, 1994; see also our discussion in the last paragraph of section 4.4).

This flare had two peaks in the time profile recorded by the soft X-ray spectrometer, one at 18 August at 00:00 UT and the other about 5 minutes later (cf., Figure 135). The soft X-ray telescope observations show four bright patches at the flare onset on 17 August at 23:58 UT (Figure 136). We number these points 1, 2, 3, and 4, starting from the northeast. First one loop appeared, connecting points 2 and 3 at about 18 August, 00:00 UT. At 00:05 UT another loop, originating from point 1, increased in brightness towards the southeast and gradually extending spatially. Four bright points in the hard X-ray telescope image appear to correspond to the points 1, 2, 3, and 4 in the soft X-ray image of 17 August, 23:58:46 UT,

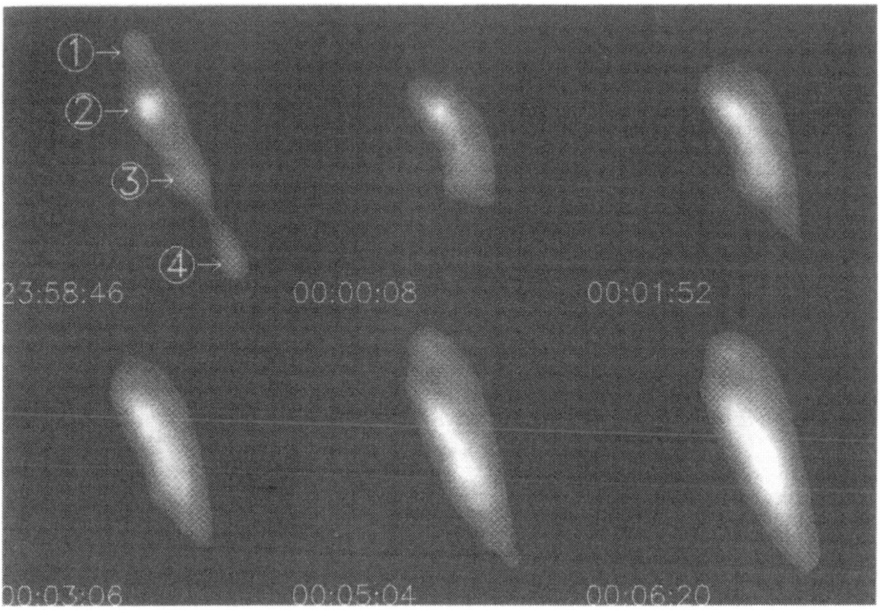

Figure 136. Soft X-ray images of the flare of 17 August 1992, 23:58 UT. For details, such as the numbering of the four patches, cf., the text. (Takahashi *et al.*,1995. From *Publ. Astron. Soc. Japan.*)

while another hard X-ray source appears to be situated between points 3 and 4 (see Figure 137). Overlaying the four points in the soft X-ray image on a magnetogram obtained by the Mitaka solar flare telescope reveals that a magnetic neutral line is located between the points 2 and 3; that the points 1 and 2 have N-polarity, while the points 3 and 4 have S-polarity (cf., Figure 138). The time profiles of the electron temperature, derived from soft X-ray emission lines in the Bragg crystal spectrometer also shows peaks on about 18 August, at about 00:00 UT and 00:05 UT.

By overlaying the soft X-ray image taken just before the flare onset (17 August, 23:56:10 UT) over the magnetogram, it is obvious that the points 2 and 3 were continuously bright, well before the flare started. During the flare a loop-like structure appeared, connecting points 2 and 3. Because these two points show opposite polarities, the structure must be a real coronal magnetic loop. Another faint loop-like structure connected to point 4 can be recognized in the (logarithmically-scaled) soft X-ray images.

The temperatures and emission measures of these four points were determined by the soft X-ray telescope filter ratio method (Hara *et al.*, 1992). Figure 139 gives a scatter diagram of the emission measure vs the electron temperature, from images observed through two filters (Be and Al). The data refer to the 9 ($= 3 \times 3$) pixel areas centered on the four points in the soft X-ray images. Correlations are found between the points 1 and 4 and between the points 2 and 3, in the decay phase of

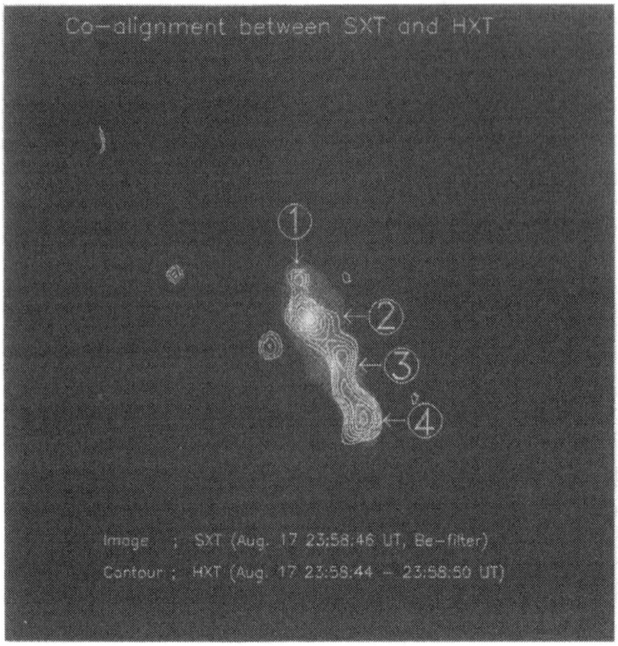

Figure 137. Flare of 17 August, 1992, 23:58 UT. Overlay of the soft X-ray image (gray-scale) over the hard X-ray image (contour). The soft X-ray image was taken on 23:58:46 UT. The hard X-ray image was obtained from 23:58:44 to 23:58:50 UT. The alignment of the two images is considered accurate to $\simeq$ 1 arc sec. (Takahashi *et al.*,1995. From *Publ. Astron. Soc. Japan.*)

the flare, which does not seem unreasonable as the two pairs of points are each footpoints of one loop.

Based on these observations, Takahashi *et al.* (1995) advanced a possible scenario for this flare: on 17 August at 23:58 UT, the first collision between the two loops took place. Non-thermal electrons were accelerated and hit the chromosphere. The four footpoints of the two loops brightened up in soft *and* hard X-ray images. A hard X-ray source (22.7–32.7 keV) was observed as well between points 3 and 4. Seven minutes later, on 18 August at 00:05 UT, the second collision of the two loops took place. The emission measure of point 1 increased rapidly. No rapid increase of the intensity of the loop connecting points 2 to 3 was observed, because the high electron density in the loop 2–3, due to the evaporated material that was already filling the loop, prevented the newly created population of non-thermal electrons from reaching the chromosphere.

A theoretical model (Chargeishvili *et al.*, 1993; Zhao *et al.*, 1993b; cf., section 4.4 and Figures 54 and 60) supports this view. This flare, which is also briefly described at the end of section 4.4, is apparently an example of a collision between two practically parallel loops (I-type coalescence). The time interval between the two peaks in the temperature-time curve can well be explained by a theoretical model (Zhao *et al.*, 1993b; Figure 60). The highest electron temperature ($\log T_e = 7.15$)

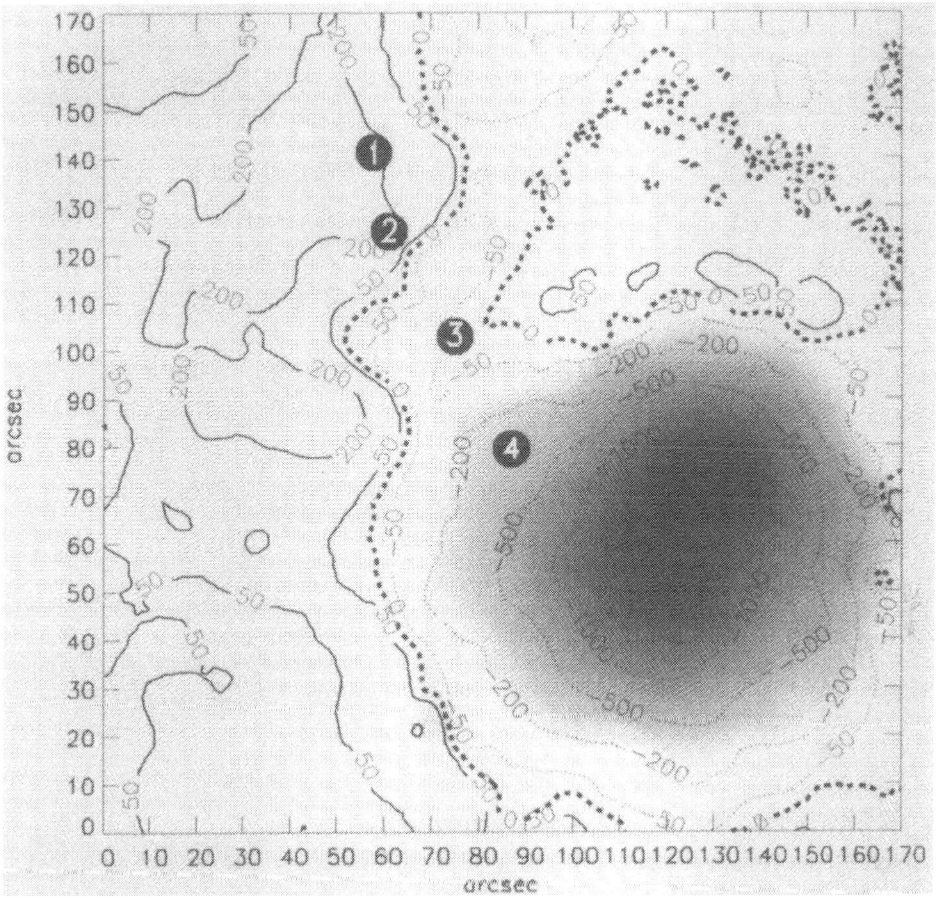

Figure 138. Flare of 17 August, 1992, 23:58 UT. Overlay of a soft X-ray image over a magnetogram. *Solid* contour: N-polarity, *dotted thin* contour: S-polarity, *dotted thick* contour: magnetic neutral line. The gray image is an optical image of a spot. The soft X-ray contours were taken from an image obtained at 23:58:46 UT. The magnetogram was obtained from 01:50:15 to 01:51:21 UT. The estimated coalignment error is $\simeq$ 2.5 arc sec. (Takahashi *et al.*, 1995. From *Publ. Astron. Soc. Japan.*)

occurred in the area between points 3 and 4 in the soft X-ray images. An additional hard X-ray source (22.7–32.7 keV) was seen in the hard X-ray image. It was located in the same region on 17 August at 23:55:44 UT. Zhao *et al.* (1993b) showed that in the region of magnetic reconnection the electrons can be accelerated to about 30 keV within a second, and that a thermal plasma of 10^8 K can be produced in collisions between two parallel loops with values of the physical parameters that agree reasonably well with the observed values. These facts imply that the region between points 3 and 4 may be the interaction region of the two-loops collision.

Enome *et al.* (1994), by comparing the *Yohkoh* data with images obtained with the Nobeyama Radioheliograph, observed a good alignment of radio, soft X-ray, and hard X-ray images of the sources in the impulsive phase of the flare.

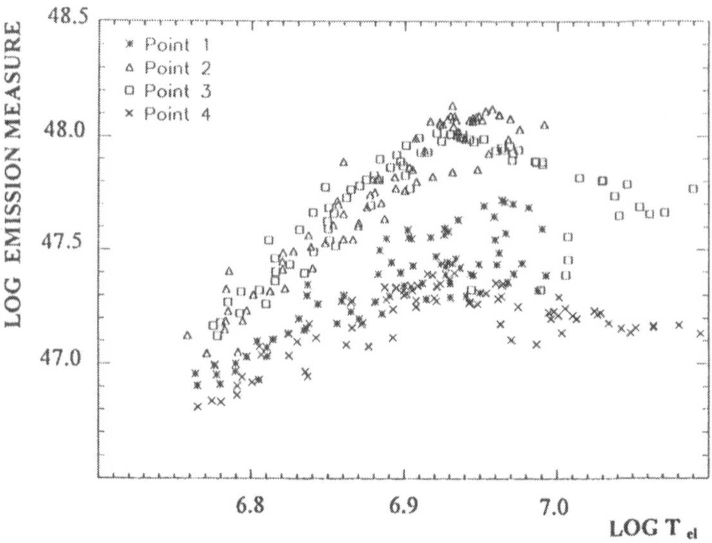

Figure 139. Flare of 17 August, 1992, 23:58 UT. Scatter diagram of the electron temperature vs the emission measure in the four areas shown in Figures 136 and 137. The data presented here were derived from the intensity ratios of *Yohkoh* soft X-ray images observed through the Al 12 μm and Be 119 μm filters respectively. Asterisks indicate data for point 1, triangles for point 2, squares for point 3 and crosses for point 4. (Takahashi *et al.*, 1995. From *Publ. Astron. Soc. Japan.*)

6.5.4. *The flare of 9 December, 1992, 10:40 UT*

A simple flare (9 December, 1992, 10:40 UT), observed by the soft and hard X-ray telescopes aboard *Yohkoh*, is remarkable in that two loops, that later merged into one, were clearly seen in the preflare stage (Inda-Koide, 1994; Inda-Koide *et al.*, 1995). This flare (GOES C2.7 class) occurred near the eastern limb ($\simeq$ E60) in active region NOAA 7363. The evolution of this active region before the flare was monitored in soft X-rays with a frequency of typically one per minute for each of the analysis filters. Soft X-ray images from the preflare to the onset of the flare are shown in Figure 140. While only one loop was seen before $\simeq$10:29:21 UT (Figure 140(c)), a second loop appeared $\simeq$7 arc sec to the east of the first, about ten minutes before flare onset, as clearly seen in the image taken at 10:37:33 UT (Figure 140(f)). The loops were approximately parallel to each other. It is important to note that in *Yohkoh* soft X-ray images the two loops were seen approaching each other already before the onset of the flare. It also appeared that the loops were not symmetric: their current-strengths were different. At $\simeq$10:38:40 UT they were almost equally bright. Slightly later, the loops seem to have collided, or to have interacted with each other, as appears in the 10:40:49 UT image (Figure 140(g)). The region of interaction was located near the top (or, more rigorously stated, slightly south of the apex) of the two loops. Finally, a remarkable brightening occurred at the two footpoints of the merging loop in the 10:41:53 UT image

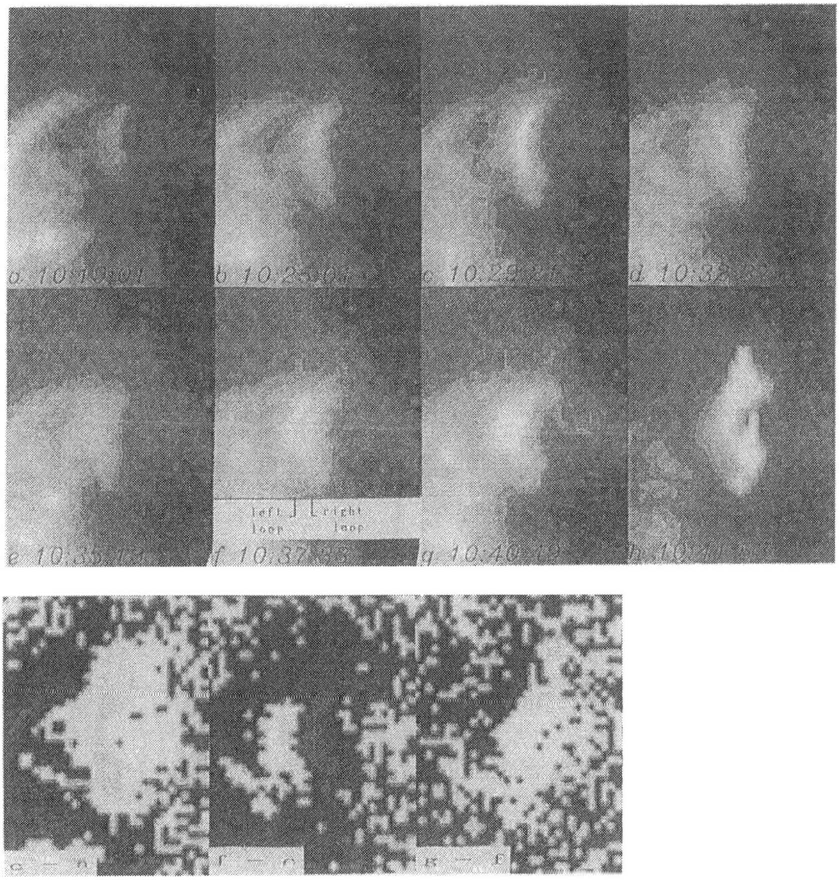

Figure 140. (*1st and 2nd rows*, a–h): soft X-ray images of the event of 9 December, 1992 in its preflare stage. The area of each image is 78.4 × 102.8 arc sec squared. The angular resolution is 2.45 arc sec. Solar north is up and east is to the left. Image (h) (second row; right below) is contaminated by CCD-saturation at the footpoints. (*3rd row*): subtractions of the images shown in the 1st and 2nd rows. The white areas shows pixels with an increased relative flux, whereas black indicates a decrease. (Inda-Koide *et al.*, 1995. From *Publ. Astron. Soc. Japan.*)

(Figure 140(h)). The temperature of the interacting region was derived from the 10:41:53 UT data and appears to be $\simeq 5 \times 10^6$ K.

The hard X-ray time profiles in the 14–23 keV and 23–33 keV bands are shown in Figure 141. Figure 142 shows corresponding hard X-ray images during the flare in the two energy bands. In the lower of the two energy bands the source was elongated throughout the flare, suggesting a trace of a magnetic loop. In the preflare phase (A), the brightest region tended to be located in its central part, probably at the loop-top. At the time of the first peak of the impulsive phase (B), the brightest region shifted northwards, and during the second peak (C) the whole loop was equally bright. The loop-top became again the brightest part of the flare in

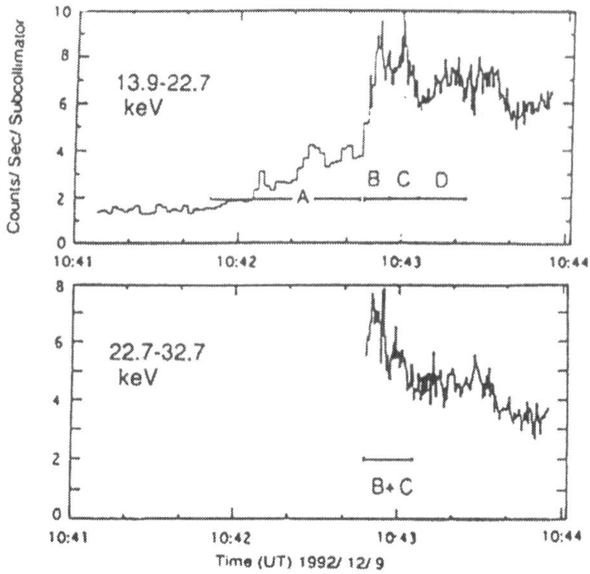

Figure 141. Hard X-ray time profiles of the flare of 9 December, 1992 in the 13.9–22.7 keV and 22.7–32.7 keV energy bands. Capitals A–D label the time intervals during which photon counts were accumulated for synthesizing the images shown in figure 142. (Inda-Koide *et al.*, 1995. From *Publ. Astron. Soc. Japan.*)

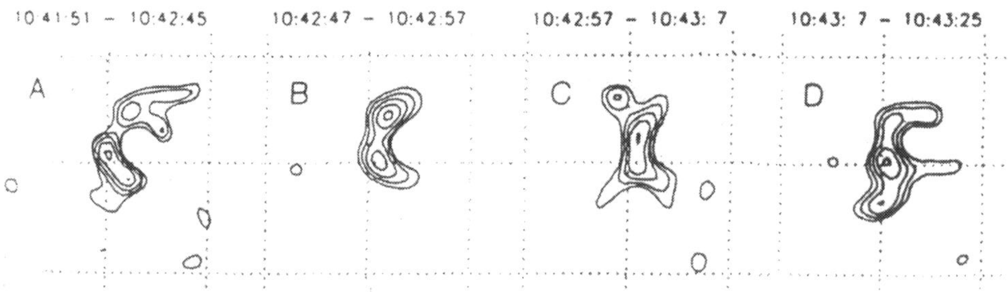

Figure 142. Hard X-ray images of the flare of 9 December, 1992. The data-accumulation intervals are given at the top of each frame. Capitals A–D refer to the time intervals marked in Figure 141. The contours are at 10.0, 17.3, 30.0, 52.0, and 90.0% of top intensity. North is up and east is to the left. (Inda-Koide *et al.*, 1995. From *Publ. Astron. Soc. Japan.*)

the gradual phase (after C). On the other hand, in the higher energy band the hard X-ray emission originated dominantly in two compact sources, separated from each other by $\simeq$30 arc sec, or 22 000 km. It was also found, from a magnetogram taken at Okayama Observatory, that the two sources were located at the magnetically opposite sides of a magnetic neutral line. Hence, the two hard X-ray sources seem to have been the two footpoints of a magnetic loop (or loops).

A probable scenario of this flare (Inda-Koide *et al.*, 1995) is as follows (Figure 143): the soft X-ray images taken during the preflare and onset stages of the flare

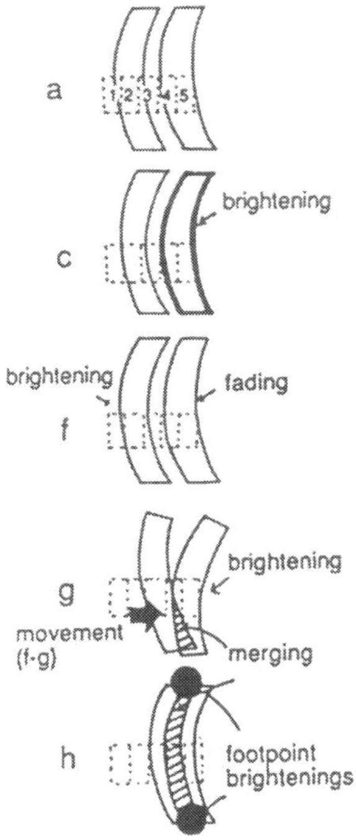

Figure 143. Schematic picture of the two-loops behavior shown by the soft X-ray images. (Inda-Koide *et al.*, 1995. From *Publ. Astron. Soc. Japan.*)

suggest that the two loops interacted, which triggered the flare's energy release. In particular the images after 10:40:49 UT show a single loop, suggesting that the two loops have coalesced or merged into one. This coalescence of the two loops took place about 2 minutes prior to the onset of the hard X-ray emission. As seen in Figure 140, the loops that coalesced several minutes later are parallel to each other. According to the classification scheme as handled in this review, this event is an I-type interaction, associated with partial reconnection. The short time scale of the interaction process, being only 2 minutes, is in accordance with the explosive reconnection model (Sakai and Ohsawa, 1987), because that model can explain rapid magnetic reconnection in a time span of the order of the Alfvén transit time.

To verify the above scenario, Fushiki and Sakai (1995c) and Sakai and Fushiki (1995b) tried to simulate the flare. To that end they performed a 3-D MHD simulation of the collision between two current-loops with different current-intensities. Initially, a strong current-loop is located parallel to a weaker current-loop with a velocity perturbation, that approaches the strong one. Figure 144 shows the time-development of a vector plot of the magnetic field produced by the weak (*upper*) and strong (*lower*) current-loops. During the approach of the weak current-loop to the stronger one, which remains in its initial position, a new current-sheet is induced between the two loops, as shown in Figure 145. Figure 146 gives plots of the resulting density and temperature surfaces. The temperature obtains its max-

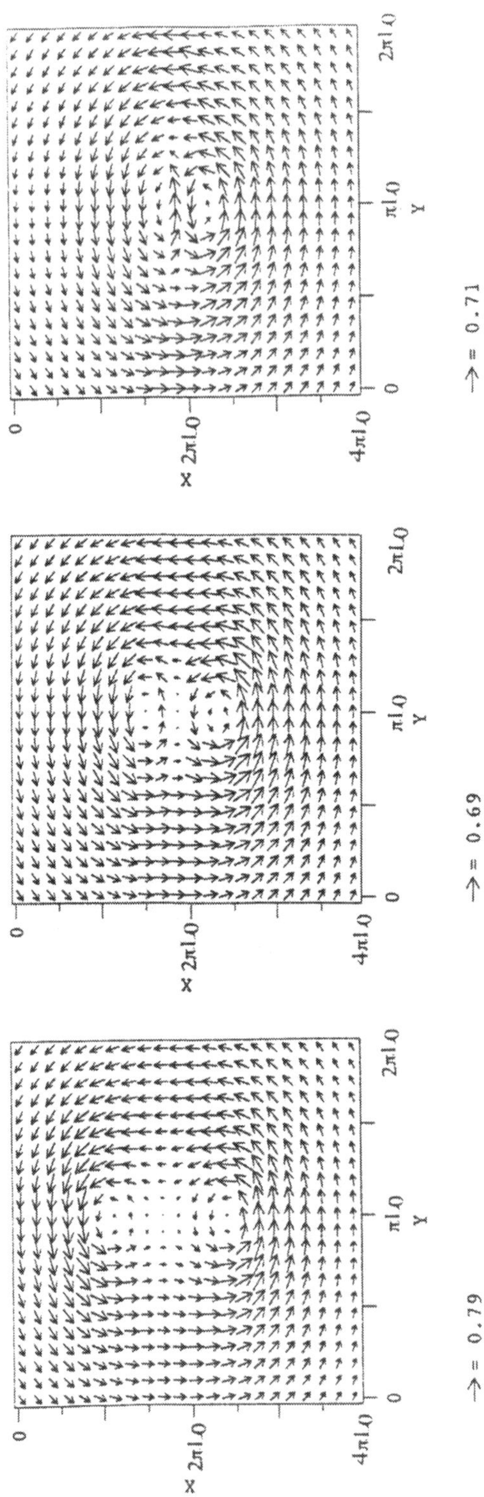

Figure 144. Simulation of the flare of 9 December 1992. Vector plot of the (B_x, B_y) magnetic field produced by weak (*above*) and strong (*below*) current-loops at the level $z = 2\pi L_0$: $t = 0$ (*left*), $3\tau_A$ (*middle*), and $5\tau_A$ (*right*). (Sakai and Fushiki, 1995b.)

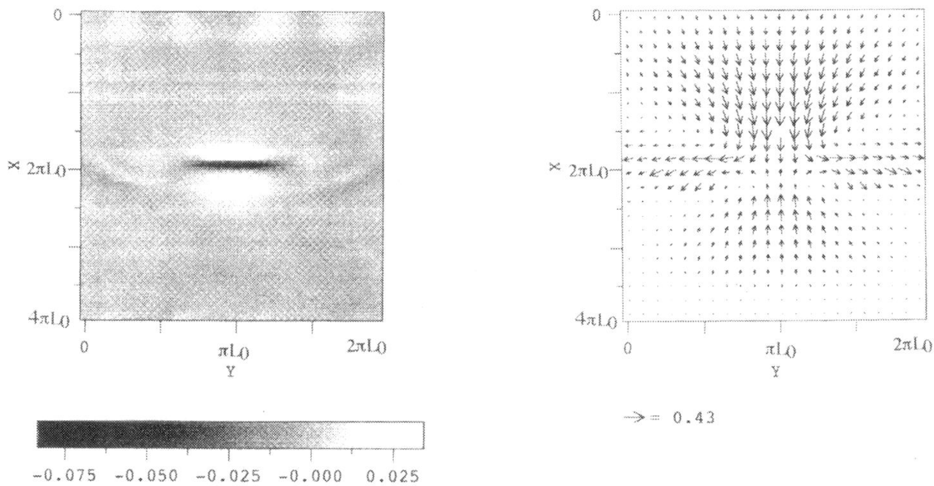

Figure 145. Simulation of the flare of 9 December 1992. *Left*: current-sheet (black region) induced between weak and strong current-loops (white region) at the level $z = 2\pi L_0$ at $t = 5\tau_A$. *Right*: vector plot of the (v_x,v_y) velocity field near the induced current-sheet at $z = 2\pi L_0$ at $t = 5\tau_A$. (Sakai and Fushiki, 1995b.)

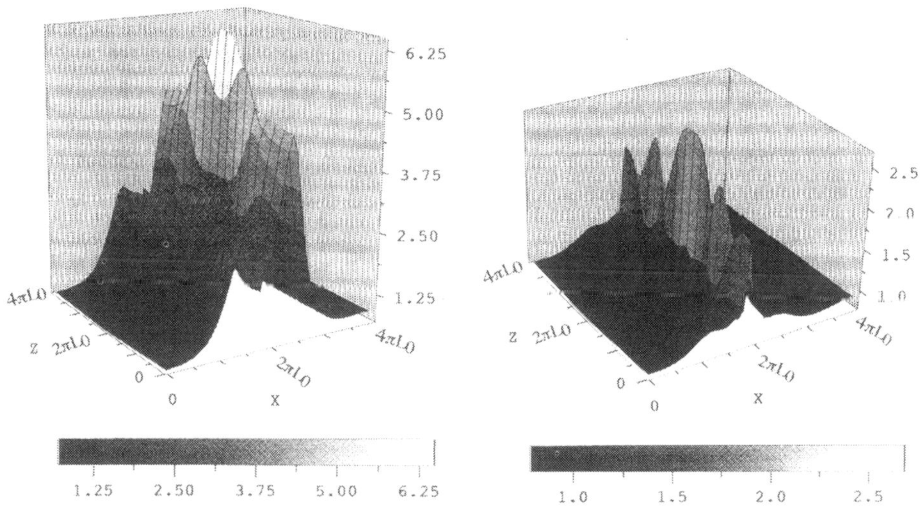

Figure 146. Simulation of the flare of 9 December 1992. *Left*: density surface plot in the x, z plane at $y = \pi L_0$. *Right*: temperature surface plot in the x, z plane at $y = \pi L_0$. The temperature obtains its maximum value (2.6 times the background) near the center of the induced current-sheet, where partial magnetic reconnection takes place. (Sakai and Fushiki, 1995b.)

imum value near the center of the induced current-sheet, where partial magnetic reconnection takes place. The temperature appears to be enhanced by about a factor 2.6 times the background value, which factor is indeed observed in the above flare.

6.6. Single-Loop Flares

6.6.1. *Introduction*

These flares are due to instabilities in singly occurring loops. The instabilities may either be due to interactions inside the loop, such as fluxthread interaction, or else be provoked by forces from outside such as emerging flux, which deforms the magnetic field system, thus leading to instability of the loop. This latter case may be considered a case of 'hidden' two-loops interaction. Any kind of instability, also the first case, needs a force from outside, because fluxthread interaction can only occur after events like footpoint motion or rotation or shearing motions in the footpoint area (cf., our sections 3.1 and 3.2).

We place the single-loop flares at this place in this section, because we consider them an elementary case of the class of confined/impulsive flares (Švestka, 1989). They have some seeming similarity to, but are certainly not identical with the cusp-type flares, which are discussed in section 6.7. The cusp-type flares are typically eruptive/dynamic flares.

6.6.2. *The Flare of 15 November 1991*

This flare deserves a prominent place in literature and may become a 'classic' one because of its role in the advancement of knowledge about flares. Sakao *et al.* (1992) found for this flare that the periods of emission of the Impulsive Phase Bursts (which they called spikes) corresponded to the period of visibility of the footpoints, while during the valleys in the intensity-time profile between the spikes, the loop was about equally bright all over. This observation definitely identifies the footpoints as the sources of the Impulsive Phase Bursts (cf., also Figures 7 and 10 in Section 2 of this review).

A study of the loop structure (Figure 8) shows a *precursor*, an *impulsive* and a *gradual phase*.

This flare was also studied by Canfield *et al.* (1992) and also for their results it deserves a prominent place in the literature on solar flares. Canfield *et al.* (1992) found that the loop was situated over the magnetic inversion line, with footpoints on either side of the line: a standard case of a single loop flare. The locations of Hα redshift, observed in high resolution spectra are close to those of the footpoints seen in soft X-ray images obtained by *Yohkoh*. The hard X-ray bursts appeared to be located at the edges of the sites of high vertical current-density. They do not coincide with these sites (Figures 11 and 12, and the corresponding text in Section 2).

6.6.3. *The Single-Loop Flare of 7 December, 1991*

A number of single-loop flares has been studied by Takakura *et al.* (1993). Two of them are described in this section and the next. The flare of 9 December, 1991 (GOES classification C6.4) had a fairly simple intensity-time profile with a main

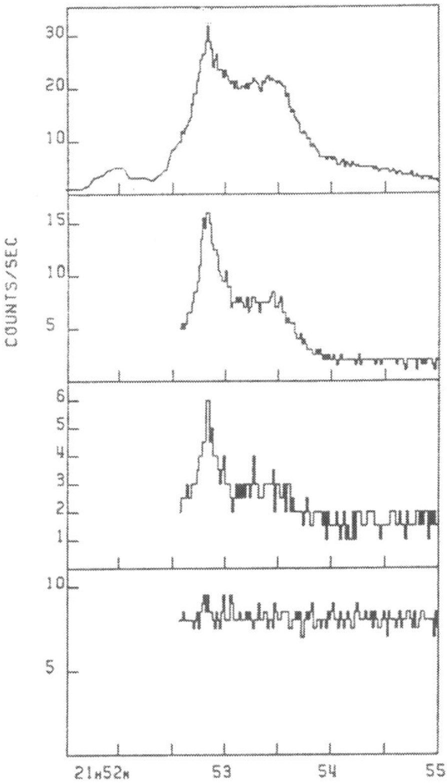

Figure 147. Time profiles of the X-ray burst on 7 December, 1991, in four energy bands. Energy ranges from *above* to *below* are: 13.9, 22.7, 32.7, 52.7, 92.8 keV. (Takakura *et al.*, 1993. From *Publ. Astron. Soc. Japan.*)

impulsive phase burst at 21:52:50 UT. It had a halfwidth of $\simeq$10 s in recordings at $\simeq$40 keV (Figure 147).

The hard X-ray images (22.7–32.7 keV; Figure 148 (left)) show that before the burst maximum the emission came from one point, in between the two sources of emission (the footpoints) that were visible at the time of maximum burst intensity. A comparison with magnetographic observations (Figure 148 (right)) shows that the flare is of the single-loop type and that the two X-ray bright areas at burst maximum are the two footpoints situated in regions of opposite magnetic polarities. Hence, the single peak that appeared before burst maximum must have been emitted from the top of the loop. A more in-depth study demonstrates that from that top the X-ray brightness spread to the footpoints during the time preceding the X-ray maximum. The speed of progression of the brightness front was 8000 km s^{-1}. Also after maximum intensity the brightest part of the loop was again at the top. The temperature at the loop-top was 5.8×10^6 K and the electron number density was 5×10^9 cm^{-3}.

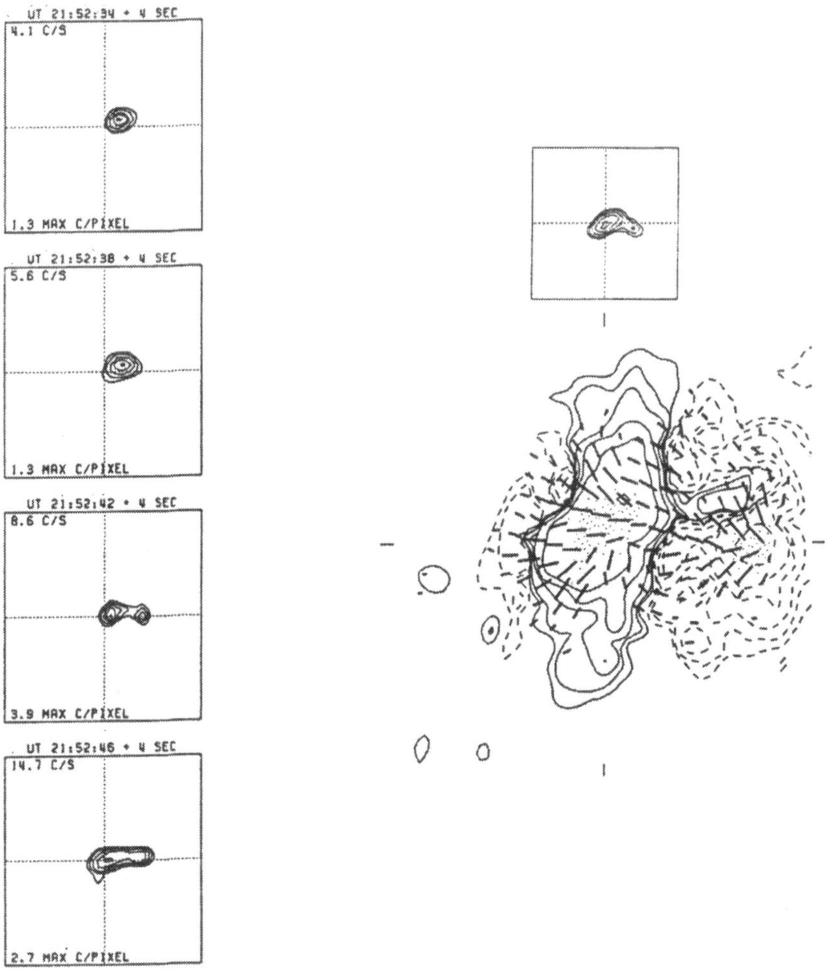

Figure 148. Left: hard X-ray (22.7-32.7 keV) contour maps of the burst on 7 December 1991; cf., also Figure 147. The starting time, and the integration time of the map are given with each panel. *Right*: vector magnetogram (*below*) compared with an X-ray map ((*upper panel*). The spatial scales in both maps are equal. For the sake of orientation the center of the lower right image is indicated by the four short bars. (Takakura *et al.*, 1993. From *Publ. Astron. Soc. Japan.*)

6.6.4. *The Flare of 21 February, 1992*

Also this flare (Tsuneta, 1996) seems to belong to the class of single-loop flares, but it may appear to be a kind of transition to the cusp-type flares, described in the next section. Energy input occurred in an X-type interaction point, situated 8 to 18×10^4 km above the top of he flare-loop, from where fast ($\simeq$800 km s^{-1}) cooling plasma streamed out.

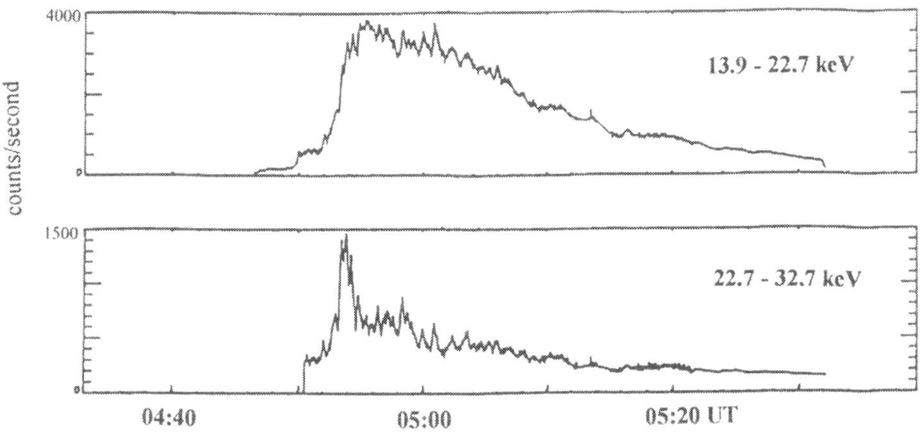

Figure 149. X-ray time profile of the flare of 2 December, 1991, in two energy bands between 14 and 33 keV. The time profiles were obtained from observations with *Yohkoh*'s Hard X-ray Telescope. (Tsuneta, 1993. From *APS Conf. Series.*)

6.7. Cusp-Type Flares

6.7.1. *Introduction*

A new type of flare that seems to be a basic form of the eruptive/dynamic ones, and that therefore may appear to be a key case in the process of understanding the flare mechanism is the *cusp-type flare*, discovered in *Yohkoh* observations. These flares are characterized by a cusp-type structure at the top of a loop or at the tops of an arcade of loops. The cusp is the hottest part (Acton *et al.*, 1992). These hot regions form at flare onset or briefly after it. They persist well, even into the decay phase. In some cases the cusps are not bigger than a few times 1000 km. The hot plasma (temperatures > 20 MK) that appears in the cusps in the early phase of the flare must be heated *in situ*, which suggests reconnection at the top of the loop.

Observations of single-loop cusp-type flares were described by Tsuneta *et al.* (1992) and Tsuneta (1993), while Takakura *et al.* (1993) and Masuda *et al.* (1994) observed a related case. Below, we discuss the flares of 2 December, 1991 and of 13 January, 1992.

6.7.2. *The Flare of 2 December, 1991*

This flare (Tsuneta, 1993) had an X-ray time profile (Figure 149) of the standard 'impulsive-gradual' type. The impulsive phase lasted from about 04:50 to 04:56 UT with one large impulsive burst complex at 04:54 UT. It consisted of at least three Impulsive Phase Bursts. The gradual phase lasted from $\simeq$ 04:50 till at least 05:30 UT. In the hour before the flare an overlying loop was situated above a compact loop that would flare later. The overlying loop was situated at a height of $\simeq$35 000 km; the flare loop at $\simeq$15 000 km. During the precursor phase, the soft X-ray brightness of the compact loop increased by about a factor ten in a 20 minute

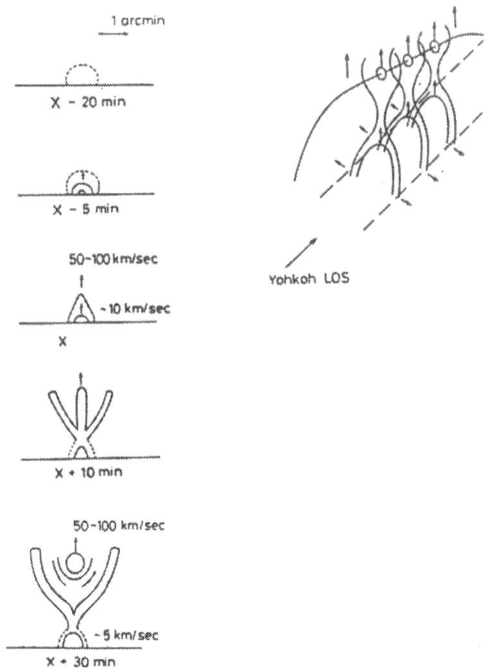

Figure 150. Left: proposed evolution of the flare of 2 December, 1991. X is the starting time of the hard X-ray flare (4:50 UT). *Right*: magnetic structure of the flare (LOS = line of sight). (Tsuneta, 1993. From *APS Conf. Series.*)

period before the start of the impulsive phase. Above the compact bright loop (the flare) and below the overlying loop a small new loop originated, about ten minutes before the main impulsive bursts ($\simeq$04:44 UT). It rose upward with a speed of 50 to 100 km s^{-1}. It continued to rise during the impulsive phase of the flare. While the flaring loop increased strongly in brightness, the rising loop evolved into a *large-scale circular-shaped blob or plasmoid* (Tsuneta, 1993) above the flare loop during the peak and subsequent decay phases of the flare (05:00 to 05:10 UT).

A comparison with magnetograms of the active region allows for a determination of a scenario of the evolution of the field line system (Figure 150). The event is assumed to have started with the appearance of a rapidly expanding loop, some 5 minutes before the flare. This was followed by the start of the hard X-ray flaring period of the flare loop. The first mentioned loop evolved into the rising plasmoid that became visible in the flare's gradual (decay) phase. What was seen as a rising blob may have been this loop, seen along its axis. It is assumed that this rising loop stretched the closed field line system as shown in the right-hand frame of Figure 150(b), and created a neutral sheet beneath it. This neutral sheet assumed about the same length as the flare loop. The upward motion of the sheet and 'blob' (actually a loop) continued with the same speed (50 to 100 km s^{-1}) but the flare loop's upward motion slowed down and came gradually to a standstill.

6.7.3. *The Remarkable 'Loop-Top' Flare of 13 January, 1992*

This flare was a limb flare, GOES classification M2.0, first described by Takaku *et al.* (1993) and thereafter by Masuda *et al.* (1994, 1995). It presented a simple intensity-time profile. Maximum intensity was at 17:28:30 UT. In energies > 20 keV it emitted one single burst with a half-width of $\simeq$ 1 minute (Figure 151). X-ray surface contours are presented in Figure 152. A comparison with magnetic maps is difficult because the limb structure had to be compared with magnetograms observed 2.6 days earlier. Yet, the comparison suggests that the loop footpoints are situated in regions of opposite polarities. The X-ray observations in 14–23 keV at 17:26:52–17:27:40 UT (i.e., on the rising part of the main burst) showed a loop extending over the limb with two clear footpoints (Figure 152). Observations in a higher energy range (33–53 keV) showed only the footpoints and a bright patch close to the apex of the loop, *but not coinciding with it.* At the time of maximum X-ray brightness this bright patch near the loop-top, as well as the footpoints, became prominently visible. The vertical distance of the patch to the loop-top was about 10 arc sec, $\approx$7000 km.

Several proposals have been advanced for explaining this flare. We describe a few of them. Masuda *et al.* (1994) and Shibata *et al.* (1995) proposed a scenario (Figure 153 (left)) in which reconnection occurred in an X-point, situated above the flare loop. From the point of reconnection a plasmoid was ejected downward and the interaction between the plasmoid and the loop, which happened near the apex of the loop, caused acceleration of electrons that subsequently moved downward along the loops. It is also expected that from the point of reconnection a plasmoid should have been ejected upward: Shibata et al. (1995) found soft X-ray erupting features in eight impulsive limb flares, whereas there were only three flares with hard X-ray sources that were situated well above (5–10 arc sec) the loop-top.

These ideas have been modeled quantitatively in a 3-D resistive MHD computer simulation by Sakai *et al.* (1995) on the basis of the model presented in Figure 75 of this review. As shown there (section 4.7) it appears that upon the approach of a plasmoid to a force-free current-loop a secondary current is produced in front of the plasmoid, until it finally coalesces with the force-free current-loop. The consequent magnetic reconnection, in turn, causes particle acceleration, which may explain the downward electron stream and footpoint bombardment.

The particle acceleration process for this case, viz. a collision between a plasma cloud and a current-loop is described in section 5.5. The predicted broken power law for the energies of the hard X-rays emitting electrons in the loop is shown in Figure 127. The two populations of electrons represented by these two lines correspond with the soft and hard X-ray sources in the loop as observed by Masuda *et al.* (1994). On the other hand, even though the predicted energy-distribution of the electrons in the cloud does not present an unambiguous broken spectrum in Figure 128, the more energetic part of the electron population should presumably correspond to the hard X-ray sources above the loop apex, as found by Masuda *et al.* (1994). Since the high-energy electron populations in the loop and the cloud

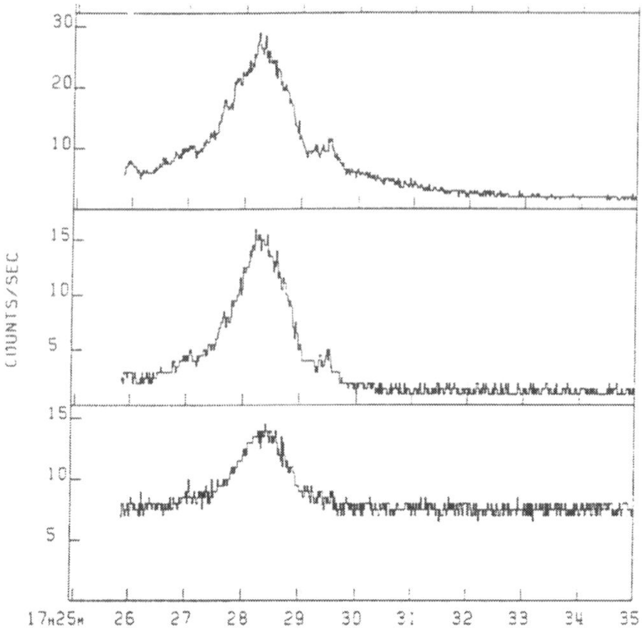

Figure 151. Time-profiles of the X-ray burst in the cusp-type flare of 13 January, 1992. Abscissa: UT. The energy band boundaries are from *above* to *below*: 14, 23, 33, 53 keV. (Takakura *et al.*, 1993. From *Publ. Astron. Soc. Japan.*)

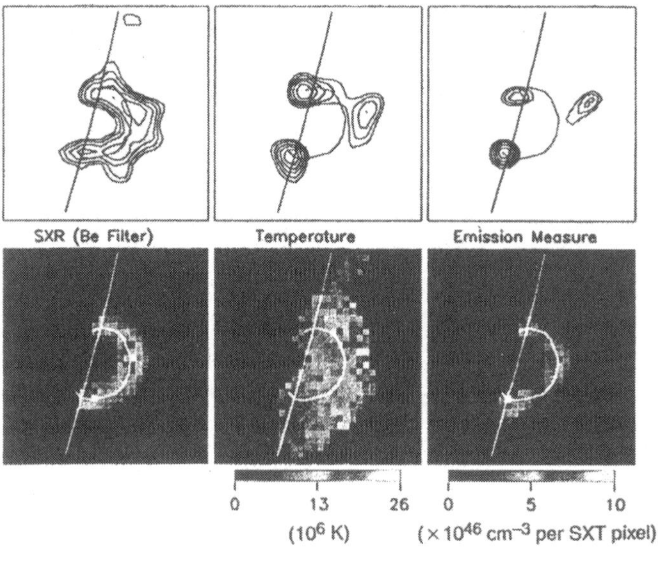

Figure 152. Upper row: hard X-ray images of the flare of 13 January, 1992, in three energy bands (14, 23, 33, 53 keV). *Lower row, from left to right*: soft X-ray images taken at 17:27:01 UT; electron temperature map, and emission measure map. Field of view: 5.7 × 10^4 km squared. North is at the top, west to the right.(Masuda *et al.*, 1994. From *Nature.*)

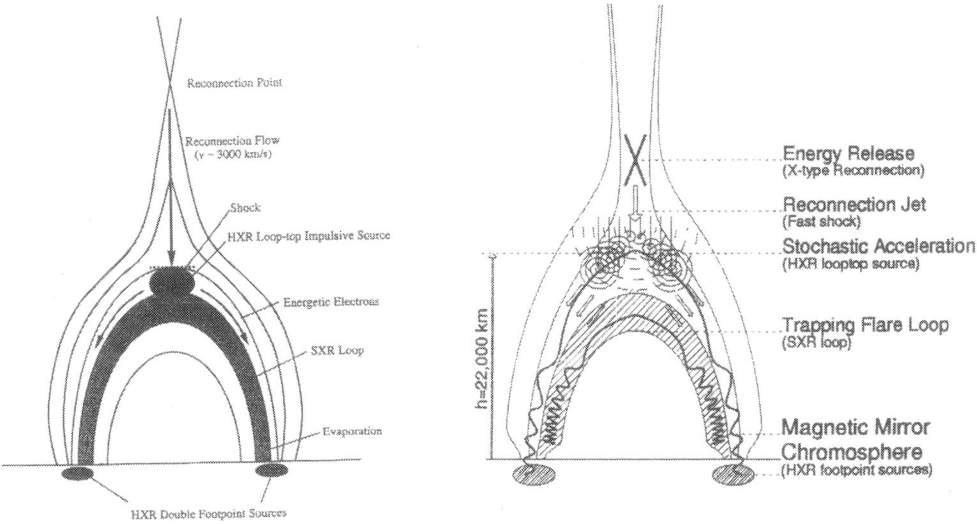

Figure 153. Left: model of the flare of 13 January, 1992. (Masuda *et al.*, 1994. From *Nature*). *Right*: alternative model. (Aschwanden *et al.*, 1995a. From *Astrophys. J.*)

are almost identical, with the same maximum energies, the combined electron populations in the simulation should be capable of producing both soft and hard X rays with the observed energies, that are radiated from approximately the predicted regions.

A different approach to the explanation of this flare was advanced by Aschwanden, Hudson and Schwartz (1995a). The new element is that they found that the hard X-rays are delayed by 40 to 220 ms in the low energy bands, with respect to the higher energy bands. Much larger delays (2 to 6 s) of opposite sign apply to the hard X-ray envelope. They ascribe these time delays to propagation time differences, and in this line of thought the calculated position of the *source* of the hard X-ray bursts must be at a height of 22000 $\pm$ 6000 km, which coincides with the loop-top source, as identified by Masuda *et al.* (1994, 1995). Hence, the loop-top emission is in this model identified with Bremsstrahlung of non-thermal electrons. The plasmoid, detected by Shibata *et al.* (1995), is in Aschwanden *et al.*'s picture related to a fast shock. The observed delay of the smoothly varying flux is explained by electrons, trapped in the flare-loop before precipitating, whereafter thick-target emission is produced in the footpoints. A model is shown in Figure 153 (right-hand diagram). The two proposed models are similar in so far as they assume reconnection in the X-point, followed by a downward directed shock. They differ in their after-effects. Which of the two gives the best representation may be decided eventually by realistic fully 3-D model calculations.

Tsuneta (1995) proposed a model in which a fast downflow from the site of reconnection collides with the loop. The resulting small-scale time-dependent shearing motions at the loop-top are expected to drive oppositely directed field-alligned current-tubes. These field-alligned currents then generate a voltage drop ($\simeq$ 100 kV) along the magnetic fields, which in turn generates runaway electrons.

6.8. Eruptive/Dynamic Flares

6.8.1. *Signatures and Consequences of Loop Interactions*

Since its launch *Yohkoh* has revealed many new aspects of solar coronal physics. Some of the results from the initial period of observations have been reviewed by Acton *et al.* (1992), Uchida (1993), Tsuneta and Lemen (1993), Tsuneta (1993a,b), and Kosugi (1993). A series of letter-papers entitled 'Initial Results from Yohkoh' was published as a special issue of the *Publ. Astron. Soc. Japan* **44**, No. 5, 1992. One of the most spectacular findings by the soft X-ray telescope in *Yohkoh* is that the X-ray corona can frequently drastically change its topology through magnetic field reconnection. Sometimes, a large-scale restructuring (Gigantic Arcade Formation) of coronal magnetic fields takes place in association with the disappearance of filaments in an event called 'propagating arcade formation', in which a closed-loop arcade is formed progressively from one end to the other in an open-field area. Some flares (Tsuneta *et al.*, 1992; Tsuneta, 1993a, b), the *Long Duration Events*, show a soft X-ray emitting bright loop with a cusp at the apex of the loop, which is similar to that found in the above described arcade formation (cf., sections 2.3.4. through 2.3.6).

Related to such observation is Yurchishin's (1994) finding that the appearance of a giant arch on 24 June, 1993 was due to reconnection between a pre-existing bundle of fluxthreads and an emerging flux loop.

6.8.2. *The Flare of 21 February, 1992; the Eruptive/Dynamic Scenario*

The typical characteristics of the eruptive/dynamic flares are described in section 2.3 of this paper. The development of the complex of processes, associated with a typical eruptive/dynamic flare usually takes a considerable amount of time, which explains why such flares are often called 'gradual flares'. We prefer the designations 'eruptive or dynamic', because these names do better justice to the formidable complex of phenomena associated with these flares.

The GOES M3.2 flare of 21 February, 1992 was already mentioned in section 2.3. Here we discuss it in more detail. The X-ray intensity-time profile (Figure 154) shows the long-enduring character of the flare, which took about two hours. Important is that the beginning of the flare, at 02:45 UT, was not the real start of the event because a coronal 'helmet streamer', visible in soft X-ray, appeared many hours before the flare. It was associated with continuous restructuring of the active region as well as with its expansion. A few phases in the development of the helmet configuration are given in Figure 155. Already the day before, 20 February, about 18 UT, this activity was observed. The helmet appeared first at 20:32 UT. A large structure, overlying a bright loop became visible around 20 February, 23:50 UT. A major expansion of this overlying loop structure started 21 February, 02:32 UT,

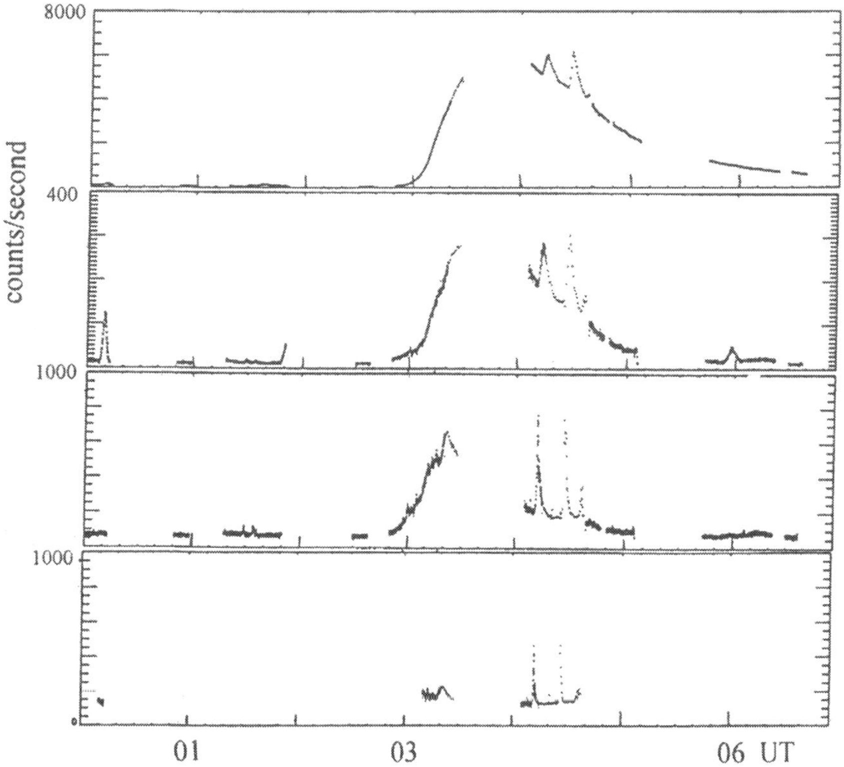

Figure 154. X-ray time profiles of the flare of 21 February, 1992. The time profiles were taken from the soft and hard X-ray telescopes aboard *Yohkoh.* From *above* to *below*: 3–15 keV, 15–40 keV, total counts of HXT sub-collimators in the low channel (13.9–22.7 keV), and in the channel 22.7–32.7 keV. (Tsuneta *et al.*, 1992. From *Publ. Astron. Soc. Japan.*)

with an upward velocity slightly above 1 km s^{-1} (Švestka *et al.*, 1995, Figure 9), from about 100 000 to 250 000 km from the limb. Type II radiobursts started at 03:07 UT. The upward motion apparently triggered the gradual start of the flare. It is suggested that a current-sheet, created by the interaction described here, is the primary source of energy of the flare, through a process of X-type or Y-type reconnection.

With this example we wish to emphasise that eruptive/dynamic flares are just one of the features associated with magnetic restructuring of an active region, related to large-scale restructuring and expansion of the region. The magnetic reconnection, occurring during this phase, gives – among other things – rise to the flare. Another effect of the high-energy phenomena, thus created, is the subsequent acceleration of *Solar Energetic Particles*, in what we call the *third phase of acceleration* (cf., Section 7 and subsection 7.4), and a consequence of the magnetic restructuring of this region is the ejection of plasma: the *Coronal Mass Ejection*.

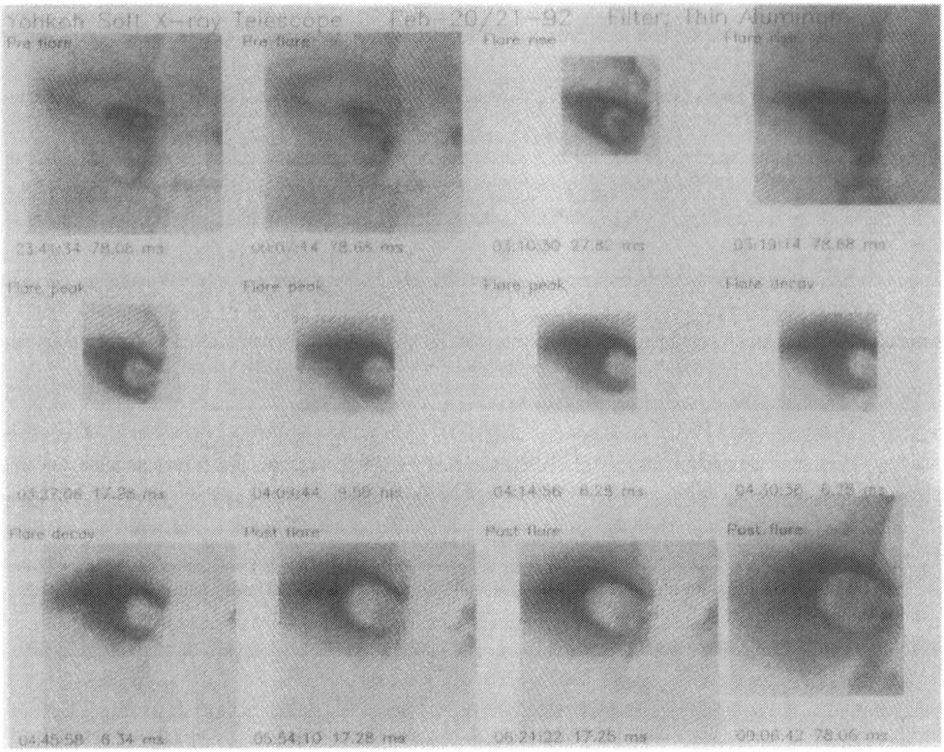

Figure 155. Pre-flare, flare, and post-flare soft X-ray images of the event of 20–21 February 1992. The resolution is 2.45 arc sec. The time of observation and the integration time are given below each image frame. North is up, and east is left. The sizes of the images are 2.6 or 5.2 arc min. (Tsuneta *et al.*, 1992a. From *Publ. Astron. Soc. Japan.*)

6.8.3. *The Dynamic Flare of 2 June, 1992*

This flare was described by Wang *et al.* (1994). There was a complicated flux-loop structure above the solar limb, along which material descended to the site where the flare occurred hours later. The peak of the low-frequency microwave emission came from the top of one bundle of Hα loops and gradually shifted to the footpoint of the loop. The microwave peak frequency increased from loop-top to the footpoints, due to the increase of the magnetic field-strength towards the footpoints. The radio-emission from the loop-top may have been due to thermal gyro-synchrotron radiation of a plasma with a temperature of 35 MK.

7. Phases of Acceleration During and After Solar Flares

7.1. Three Phases

In section 4.6.2 we mention the hypothesis of de Jager (1988) and Sakai and de Jager (1991) that there are three phases of acceleration in a fully developed eruptive/dynamic flare. Strictly speaking, a confined flare, by its very nature, can not have more than the first two phases, because the third phase implies a field line configuration such that particles energized in the first two phases can be further accelerated and escape to space. Such a configuration is not expected for a confined flare. Yet, there are exceptions: there are confined flares with a significant flux of interplanetary particle emission. We think that these particles escape by some kind of diffusion process and this hypothesis is substantiated firstly by the observation that the flux of such particles is generally small and also because the ionic composition of these particles reflects the signature of the second phase of acceleration and the subsequent outward diffusion.

Several observational indications, upon their combination, suggest that there are two early phases of acceleration in a well-developed flare, no more than seconds apart but based on very different physical mechanisms. They are followed – but mainly in a fully developed eruptive/dynamic flare – by a third phase, tens of seconds to several minutes later.

The relevant observations regarding the first two phases are:

– The acceleration times of electrons and protons to energies of the order of one MeV are of the order of a second.

– Although MeV ions are accelerated nearly simultaneously with MeV electrons, observations with high time resolution show that the ionic emission, up to energies of about 10 MeV, can occur some seconds later than that due to electrons. This requires a second phase of acceleration for which shock-acceleration is proposed.

Theoretical work done so far refers mainly to the first phase and is described in sections 4.5 and 5.3. In the first of these, acceleration during X-, Y-, or I-type coalescence was studied, and it was found that for the cases of two-loops interaction studied here, Y-type coalescence is the most promising configuration for efficient acceleration. It was predicted that it can cause prompt acceleration of protons to <10 MeV within 0.001 s, and of electrons to <5 MeV within 10^{-6} s. For X-type coalescence, theory claims that protons and electrons can be accelerated within $\simeq 1$ s to maximally 70 to 100 Mev or 100–200 MeV, respectively.

In the case of partial reconnection of kink-unstable loops (which seems a more realistic model than those described above) it seems (section 5.3) that electrons can be promptly accelerated to relativistic energies, while the problem of ion-acceleration has not been investigated. It seems though hat under exceptional conditions ions might be promptly accelerated to GeV energies (Litvinenko and Somov, 1995).

The process of shock-formation, important for the second and third phases, has not yet been studied in sufficient detail. For the case of partial reconnection during two-loops coalescence the formation of fast magnetosonic shocks was confirmed by a 3-D MHD simulation (Koide and Sakai, 1994). They found that a time interval of the order of the Alfvén time, τ_A is needed for shock formation. We will assume this value in the following. To give an impression: for $B = 30$ G, $n_e = 10^{10}$ cm^{-3}

and $l = 1000$ km, $\tau_A = 1$ s. (The shorter (subsecond) time scales found by La Rosa *et al.* (1994) and LaRosa *et al.* (1996) are based on assumed shorter path length scales.)

In the following three subsections of this Section we deal with the observational evidences for the existence of these three phases, but before doing so we outline their scenario (reference is also made to the discussion by Chupp, 1983b).

(1) Phase of *explosive coalescence and acceleration* in a d.c. electromagnetic field. During the pre-flare implosion at reconnection the magnetic field is strongly amplified by the local magnetic pinch effect. Thus, an almost one-dimensional current-sheet originates in which prompt high-energy particle acceleration can occur; the energy that can be reached depends on the configuration. This theory is supported by the observation that non-thermal emission due to electrons appears to peak at the same time in all energy bands (keV, MeV; Yoshimori *et al.*, 1992, Yoshimori, 1994) at the onset of the impulsive phase. Zarro *et al.* (1995) found that the hard X-ray emission is due to electrons that are accelerated in a d.c. electric field of 10^{-4} V cm^{-1}. This result fits with what we know of current-loops reconnection. In order that a significant fraction of ions is accelerated in a d.c. electric field, that field should be of the order of, or less than the Dreicer field (for which the entire electron population is in the runaway regime; Holman, 1995).

(2) The coalescent explosion leads to a *MHD system of shock waves* or, formulated more generally *MHD turbulence*. As suggested by Sakai and Ohsawa (1987), Ohsawa and Terashima (1993) and others, ions can be accelerated to high energies in a situation in which an electric field propagates at right angles to the **B**-field (in a spherical explosion such a situation must in any case occur somewhere). If the electric field is strong enough for ions to be trapped, they will be accelerated in a direction at right angles to **B** and **E**. LaRosa *et al.* (1996) find that in this phase electrons can be accelerated over protons to bulk energies of 25 keV. Other authors, assuming acceleration over a path of a few hundred km (which may be the characteristic pathlength in the region of coalescence) find resulting proton energies of $\simeq 100$ MeV, a value that is necessary for explaining the observed γ-ray emission. If the available path length would be longer, GeV energies could be reached. The very different predictions of energies of accelerated particles, found in literature, can chiefly be brought back to such differences on the assumed path lengths of acceleration.

It is in this configuration that selective He^3 acceleration, as well as differential acceleration of other ions seems to occur.

(3) Many authors, cf., Cliver *et al.* (1985) have found observational support for the long-standing suggestion (Wild, 1965; de Jager, 1969) that there is a later (third) phase of acceleration. Support is found in the observation of type-II and type-IV radiobursts, and in the observation of Solar Energetic Particles in interplanetary space. These particles can have obtained their high energies by *shock wave acceleration* in a field line system situated above the flaring region, and which opens towards space. This type of acceleration, hence, occurs in principle only in

eruptive/dynamic flares, although Ramaty and Murphy's (1987) may be right with their remark that there is no full separation between the cases.

In the following sections we summarize observational evidence.

7.2. Observations Relevant to the Time Differences between the First Two Phases and to Their Energies

In view of the small time difference Δt expected between the first and second phases, a determination of these values demands accurate timing. The initial observations led to the conclusions that the time differences between the production of <200 keV electrons and MeV ions are 0 ± 2 s (Forrest and Chubb, 1983) or 0 ± 1 s (Kane *et al.*, 1986) but when more precise observations became available, it was found (Figures 156 and 157) that the MeV radiation can arrive a few seconds later than those of lower energies. A value of 2 to 4 s was found for the flare of 7 June 1980 (Chupp, 1983a; Figure 156) and that of 3 June, 1982 (Forrest *et al.*, 1986). A larger Δt-value was found for the flare of 27 October, 1991: Yoshimori *et al.* (1992) found that the ionic emissions in the MeV range arrived 14 s later than the Bremsstrahlung from electrons in the 100 keV range. From Figure 157 (Kawabata *et al.*, 1994), which refers to the flare of 15 November, 1991, one finds that emissions below $\simeq$500 keV are observed a few seconds before those of higher energies.

The above enumeration shows that there is a tendency for high-energy radiation to be emitted slightly later than those at lower energies, but there are exceptions. Chupp *et al.* (1993) and Trottet *et al.* (1994) studied the flares of 9 September, 1989, 09:10 UT and of 3 June 1982, 11:42 UT, on the basis of optical, radio and high-energy emissions. They found that transient bursts of >10 MeV are emitted in the rising part of an impulsive gamma-ray burst. These two kinds of high-energy bursts are ascribed to *two different particle populations*, associated with the appearance of a new Hα feature (case of 9 September, 1989) or to acceleration in different magnetic structures (3 June, 1982). This result therefore does not conflict with the former. We think that the complicated structures of these two flares hampers obtaining clear views on elementary acceleration processes and *conclude* therefore that the scanty information so far available points to $\Delta t \approx$ 2–14 s.

The above studies also present observational data on the energies involved in the first two phases: in the first phase electrons are observed to be accelerated to energies of at least some 500 keV. Kocharov *et al.* (1994) find that in the impulsive phase of flares electrons are accelerated up to energies of 2 MeV, but that information is too crude and it does not yet allow to decide if MeV electrons are accelerated at the same time as those of 100 keV. The decision should come from γ-ray continuum observations with a high timing precision. For the time being *we assume 1 MeV for the highest observed energy of electron acceleration in the first phase*. Prompt γ-ray lines, up to 10 MeV, are observed in the *second phase*, and

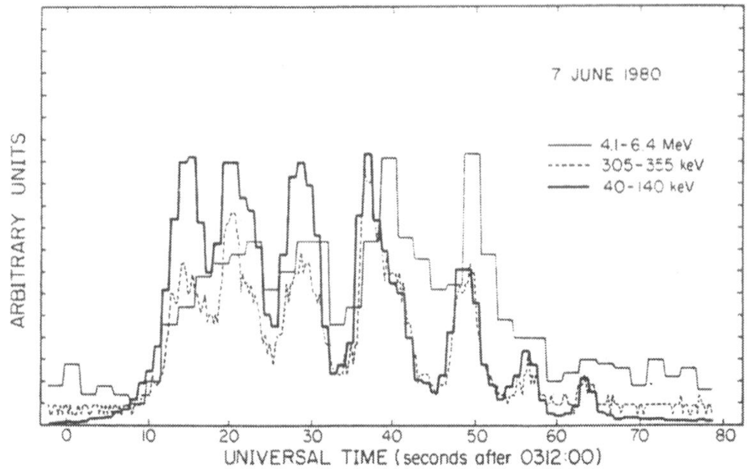

Figure 156. Time profiles of hard X-ray and gamma-ray line emissions for the flare of 7 June, 1980, showing a $\simeq 2$ s delay of the MeV radiation. (Chupp, 1984a. From *Solar Phys.*)

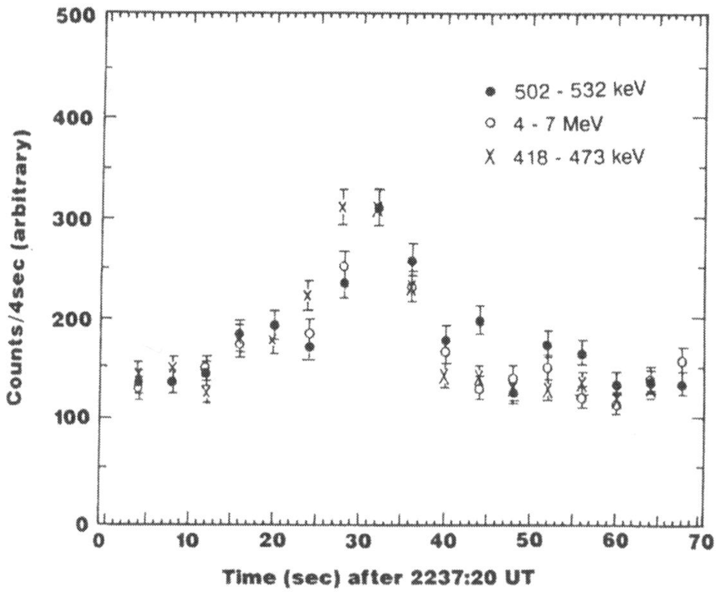

Figure 157. Time profiles of counting rates for the 502–532 keV, 4–7 MeV, and 418–473 keV band emissions in the flare of 15 November, 1991, showing a relative delay of the most energetic emissions. (Kawabata *et al.*, 1994. From *Astrophys. J.*)

this means *ionic acceleration up to energies of 50–100 MeV in the second phase.* We also deal with this question in the next section.

As to the mechanisms of acceleration, we note that Zarro *et al.* (1994) find that acceleration of soft and hard X-ray emitting particles occurs in a d.c. electric field with a strength of several times 10^{-5} V cm^{-1}, in a compact region (first phase).

Next, Ramaty and Murphy (1987) found that the γ-ray emission comes from particles trapped in closed loops - hence in a situation that could be identified with a confined flare or with the first two phases of acceleration. Finally, the injection of relativistic electron beams (type III radio bursts) is coherently modulated by either a compact source or one that has a global organisation on a time scale of seconds (Aschwanden *et al.*, 1994). This latter aspect definitely refers to the second phase.

We have to add that there are also *delayed radiations*; their delays in emission are due to other effects than those described above. White-light flares occur 1 to 2 s after the hard X-ray emission (cf., Neidig *et al.*, 1992); this effect is due to energetic particles scattered in the photosphere, and consequent backwarming (cf., section 2.2.4). The annihilation line at 511 keV needs baryon deceleration before positrons can be produced.

7.3. The Second Phase: Shock-Wave Acceleration after Coalescence

The principal sources of information on this phase are the γ-ray and neutron emissions that can be attributed to shock-wave acceleration, while data on the selective ionic (mainly He^3) acceleration inform us about the relevant wave forms. There is some information about the energy range involved: Alexander *et al.* (1994) in discussing γ-ray spectra of two flares found considerable emission above $\simeq$10 MeV and concluded that the spectra and spectral indices can be explained by introducing a cut-off in the accelerated ionic spectra at 500 to 800 MeV. Kocharov *et al.* (1993) discussed the neutron and gamma-ray emission of the flare of 24 May, 1990. There are fast and slow components of γ-rays; the former with energies up to 110 MeV, the latter in a similar energy range. Their emission is attributed to shock-wave acceleration and the decay time of the fast component ($\simeq$20 s) is thought to be an indication of the decay time of the shock-wave field that originated, we think, during the first phase of flux-tube coalescence. We also think that the slow decay time of $\simeq$ 260 s may be attributed to the shock-wave field that originated in the later (third) phase of acceleration.

We *conclude that during the second phase ions can be accelerated to energies of the order of a few hundred MeV*, with considerable scatter; there is no observational information about the energies to which electrons are accelerated during this phase.

It is in this 2nd phase that *differential ion acceleration* takes place, the most remarkable aspect being the differential He^3 acceleration. Figures 158, 159, and 160 summarized and reviewed by Kocharov and Kocharov (1984) from data by Hovestadt *et al.*, Ramaty *et al.* and Hempe (references *loc. cit.*) give first (Figure 158) average values for the enrichment factors (with regard to 'cosmical values') for ions from H to the Fe-group. All kinds of ions can be more or less enhanced, but He^3 is the most spectacular case. That there is some correlation between the He^3/H ratio and the He^4/H ratio is shown by Figure 159, while Figure 160 shows that the He^3/He^4 ratio depends somewhat on the maximum proton flux. These data, although some 10 years old, are still correct in broad outline. To this data we have

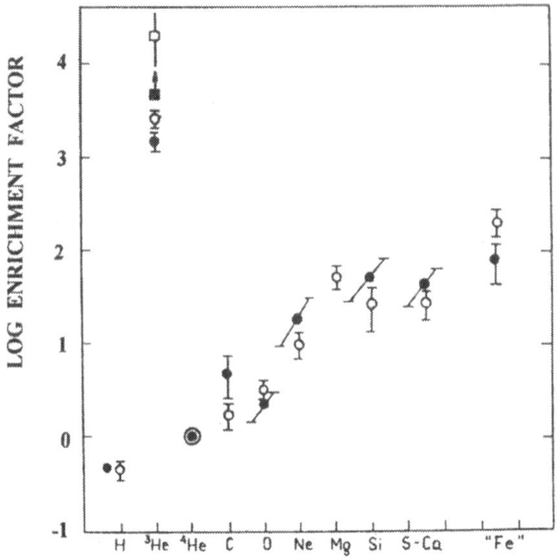

Figure 158. Mass-dependence of the enrichment factor of various ionic species relative to He^4 for the energy ≈ 1 MeV nucl^{-1} for the events of 9 May, 1974 and 30 November, 1974. The ordinate gives the 10-log of the enrichment factor, assumed unity for He^4. (Kocharov *et al.*, 1984. From *Space Sci. Rev.*)

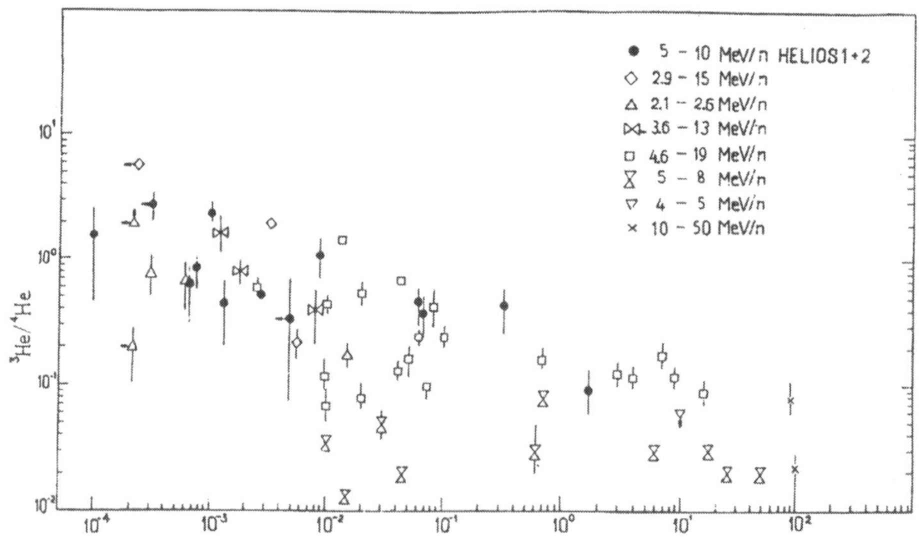

Figure 159. He^3/He^4 ratio vs maximum proton flux in the energy range 4–13 MeV. The proton flux is given per cm^2 sr MeV. (Review by Kocharov *et al.*, 1984. From *Space Sci. Rev.*)

to add, that Kallenrode and Švestka (1994) found that also the He/electron flux ratio in interplanetary space can vary greatly, by a factor of at least 15 or more.

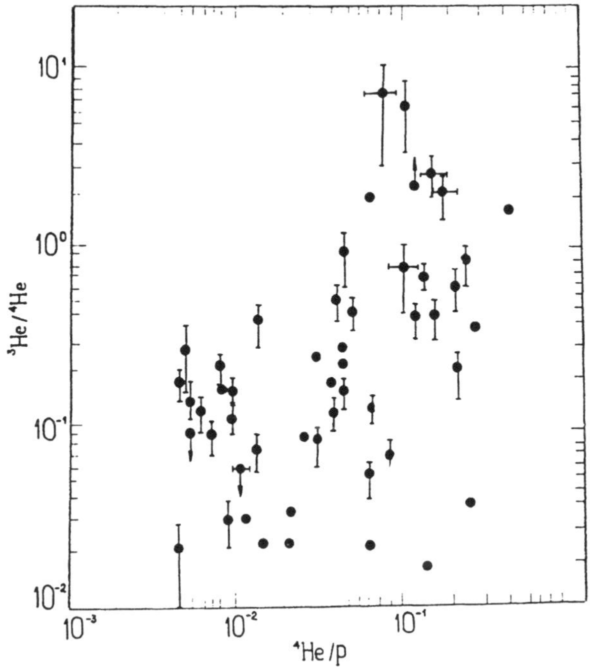

Figure 160. He^3/He^4 ratio vs the He^4/p ratio for solar cosmic rays. (Kocharov *et al.*, 1984. From *Space Sci. Rev.*)

With regard to the source of these variations in ionic and electron composition the question is where they originate: in interplanetary space or in the corona, and in the latter case, where. The electron concentration variations were discussed by Kallenrode and Švestka (1994), who concluded that these originate in the type III nuclei. In our 3-phase terminology we would therefore attribute these to the second phase of acceleration (cf., also Aschwanden *et al.*, 1994). In the study of the ionic composition an important step forward was made when Murphy *et al.* (1991) determined the chemical composition of particles from the flare of 27 April, 1981 both in space and in the flare, the latter from the γ-ray spectra. They found that ionic abundances of Mg, Si, and Fe were the same in the flare and in space, but enhanced with regard to the photosphere. This shows that the enrichment occurs in the flare environment, during our second phase of acceleration. Kocharov *et al.* (1993) studied the γ-ray spectrum of the flare of 21 June 1980; they found that the relevant particles were accelerated in a medium with $n_e = 10^{11}$ cm^{-3} and $B = 300$ G. These particles were accelerated in a few seconds in a system of shock-waves, perhaps originating from explosive current-loop coalescence (second phase). Miller *et al.* (1990) investigated stochastic Fermi-type acceleration in solar flares and concluded that these can account for the impulsive phase ions of the flare of 3 June 1982 (our 'second phase') but not for the 10–60 MeV relativistic electrons (apparently our 'third phase').

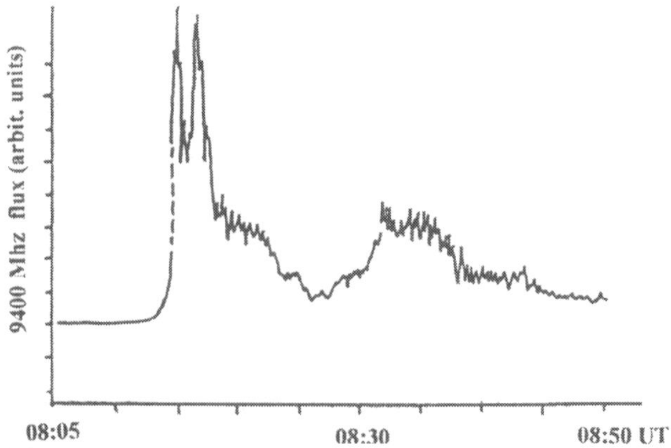

Figure 161. Radio-burst of the flare of 15 June 1991, observed at Toyokawa at 9400 MHz. (Belov *et al.*, 1996. From *Sendai Solar Terr. Phys. Proc.*)

Several theories have been advanced to explain the ionic enrichments. All of them demand the existence of a wave field. Kocharov and Kocharov (1984) advanced the theory of differential acceleration in ionic waves, in a plasma where the electron temperature is much higher than the ion temperature – a situation that is not impossible because the heating time of electrons in coalescence is much shorter than of ions; for the latter a time of a few seconds seems likely (de Jager *et al.*, 1987). In their fundamental review paper, Ramaty and Murphy (1987) treat the acceleration in what they call the 'first phase' – which is our second phase – and conclude that in that phase stochastic acceleration operating in closed loops – our confined flares – *produces the particles responsible for the bulk of the gamma-ray and neutron emission.* Miller and Ramaty (1993) propose a combination of nonlinear Landau damping and gyroresonant acceleration for this process. Ohsawa and Terashima (1993) find the solution in first order Fermi acceleration in a large magnetosonic wave at the right frequency. Ohsawa (1993), elaborating these ideas, develops a 'unified theory' of He^3 enrichment and 'prompt' (i.e., second phase) particle acceleration in non-linear magnetosonic waves with a strong B-field in which the cyclotron frequency exceeds the electron plasma frequency, and in a later study, Zhang and Ohsawa (1995) find the solution in unstable H cyclotron waves in a plasma with $T_e/T_i > 1.5$. Whatever the mechanism will turn out to be, it needs a system of magnetosonic waves, most probably of magneto-ionic shocks.

7.4. THIRD PHASE: STOCHASTIC SHOCK WAVE ACCELERATION IN A LATER STAGE

Our third phase is what Ramaty and Murphy (1987) call the second phase. They remark *l.c.* p. 263: *...there are flares for which there are more escaping than interacting particles. Here the escaping particles are probably accelerated on open field lines....The second phase operates on open field lines and produces the particles seen in planetary space.*

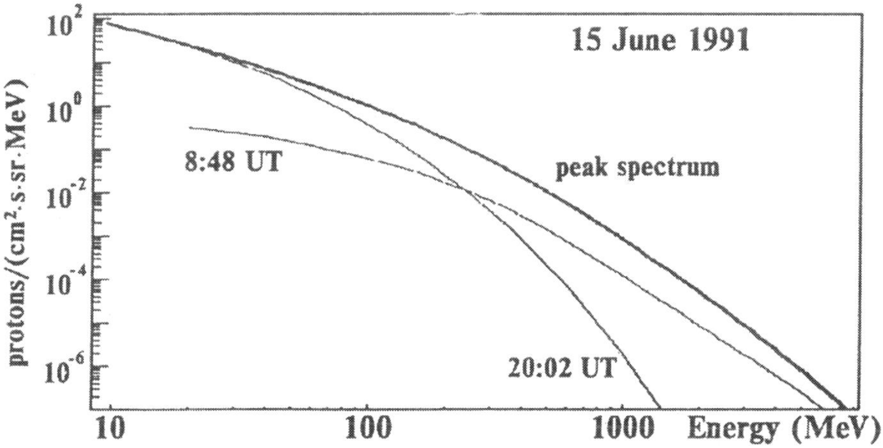

Figure 162. Differential proton spectra observed near Earth after the flare of 15 June 1991 at different times, together with the peak spectrum. (Belov et al., 1996. From *Sendai Solar Terr. Phys. Proc.*)

An illustrative case is the flare of 15 June, 1991 (Belov *et al.*, 1996). The flare had strong impulsive bursts between 08:17 to 08:23 UT (Figure 161), at which time white-light emission was observed in compact loops of about 10 arc sec size. There was an extended gradual phase, recorded on microwaves, that lasted at least till about 08:50 UT. Gamma-ray emission between energies as high as 1 to 2 MeV was observed during $\simeq$2 hours after the impulsive phase. Solar protons with energies between 0.1 and 1.6 GeV were observed in space near the Earth. Figure 162 gives the differential energy spectra at two times, early and late in the flare, and the peak spectrum.

Assuming momentary emission of the protons, followed by isotropic diffusion, the time dependence of the injection can be derived on the basis of an integral equation developed by Krimigis. The result is shown in Figure 163, which shows that the injection time becomes progressively later for higher proton energy, particularly in the energy range 100–400 MeV. In that range the injection time runs from $\simeq$08:18 to $\simeq$08:26 UT. For still higher energies the derived injection times are too uncertain for any firm conclusion. The importance of this diagram is that it demonstrates clearly the acceleration process: acceleration to higher energies takes more time than to lower values, an observation that is fully compatible with current ideas on third phase shock-wave acceleration in the corona. We also refer to Simnett (1996) who summarized reasons why energetic particles up to relativistic particles, detected in interplanetary space, should have originated in the corona.

In conclusion, we think that there is ample evidence that the third-phase acceleration occurs some time, up to minutes, after the associated flare in the shocked field-line system higher up in the corona. During this phase ions can be accelerated to above GeV energies.

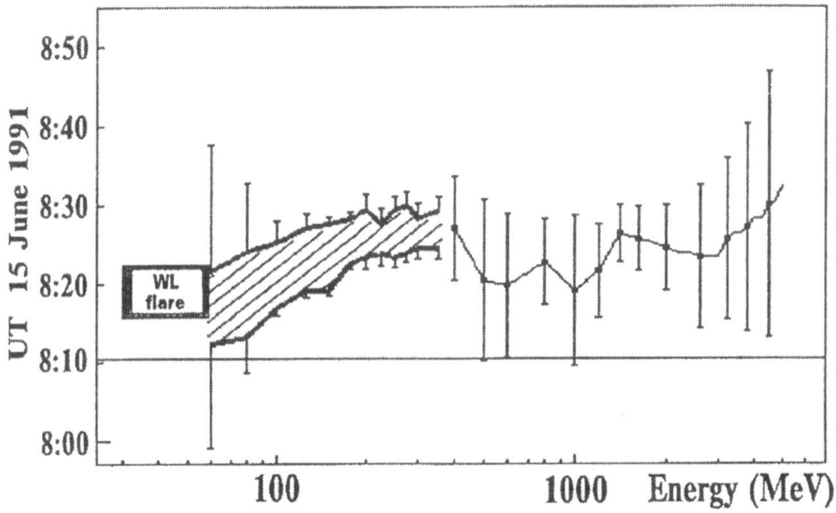

Figure 163. Time and duration of the proton injection after the flare of 15 June, 1991. In the 60–350 MeV energy range the derived injection times are presented as a hatched region, limited by the solid curves. For >350 MeV energies the injection times with r.m.s. errors are given. The horizontal line is the X-ray onset time. (Belov *et al.*, 1996. From *Sendai STP Proc.*)

8. Summaries

8.1. Flare Observations and Flare Simulations

In Section 2 we summarize the properties of solar flares, which we broadly distinguish in confined/impulsive and eruptive/dynamic flares. For brevity we call them sometimes just 'confined' or 'eruptive'. This classification is mainly a topological one and corresponds to 'closed' or 'open' field-line systems. In general, the classification appears to be quite satisfactory. There are only few cases that do not fit into this simple scheme.

In Sections 3–5 we discuss the theoretical aspects of single-loop instability and of two-loops interactions, in order to compare these results (in Section 6) with flare observations.

Section 3 discusses single current-loops. Upon footpoint rotation internal currents and counter-currents develop. In certain cases they obtain a cylindrical shell-like structure. These current-shells are unstable and desintegrate. Thus, many helical fluxthreads (current-filaments) originate. This result offers possibilities for the explanation of certain observations.

The collisional interaction of two current-loops is discussed in Section 4. We define three possible cases, dependent on the mutual positions of the interacting current-loops. We call them I-, Y-, and X-type coalescence. For these three types the results from coalescence simulations differ among other things in the time-developments of the temperature: X-type coalescence is associated with only one temperature maximum in time, while Y- and I-type coalescence rather yield pulsating temperature-time curves. Particle acceleration is most efficient in Y-type coalescence and for this case the strongest acceleration occurs near the surface of the region of interaction.

Dynamic instabilities of single loops and of interacting loops are dealt with in Section 5, partly on the basis of many-particle simulations. Three successive types of instabilities may occur: Buneman instabilities, pinching and kink instabilities. The simulations result in quantitative data on high-energy particle acceleration and on electromagnetic wave emission. Section 5 also deals with the collision of a plasma cloud with a current-loop, a phenomenon that seems to occur in the cusp-type flares.

In Section 6 we describe and classify a number of observed flares in the light of these theoretical developments. First we decribe flares due to interactions of the three types mentioned above (I, Y and X). Then follow descriptions of single-loop flares, the cusp-type flares, and finally the eruptive/dynamic flares.

We summarize our ideas on flare types and flare classification:

The three kinds of two-loops interaction, described above, can occur in confined as well as in eruptive flares. Single-loop flares are confined flares. They can be caused by single-loop instability. There are various possible scenarios for single-loop instabilities, ranging between fluxthread interaction within a current-loop and single loop instability related to emerging magnetic flux. Often, the interactions appear not to be complete or they are even only marginal: various observations demonstrate the importance of partial reconnection. It also appears that reconnection need not be a single process in a flare; many observations show that the process may be repeating and repetitive, and that it continues during some time: *ongoing reconnection.*

The cusp-type flares are apparently single-loop structures in which the instability seems to start in an X-point or X-line above the loop, which leads to opening of the field-line configuration. Therefore the cusp-type flares belong to the class of eruptive/dynamic flares.

The eruptive/dynamic flares are related to a *grand instability* in a large solar magnetic field system, also involving instabilities in loops, or in arcades of loops. Such a flare is just one of the events associated with that instability. Another element of a grand instability is the associated instability or 'disappearance' of filaments (they do not really disappear). A further element is the Coronal Mass Ejection. An eruptive/dynamic flare is therefore not the cause of the CME, but rather are flare and CME due to the more basic grand instability that leads to the flare and to mass ejection. Various causes may be considered for this instability; the most common and most probable feature is an emerging magnetic flux.

8.2. PHASES OF ACCELERATION

The various phases of acceleration are discussed from the observational point of view in Section 7, and theoretical work is treated in sections 4.5 and 5.3. We distinguish between three phases of acceleration. The first phase occurs practically instantaneously during coalescence in a d.c. electromagnetic field that is formed during coalescence. Observations show that in this phase electrons can be accelerated to energies of about 1 MeV.

Coalescence leads to the formation of MHD turbulence that may develop into a field of MHD shocks. In the second phase particles are further accelerated in this field of shock-waves. Observations show that this phase starts between less than one to some ten seconds after the first, which is the time expected to be needed for building up the shock-wave field. In this phase, ions can be accelerated up to energies of a few hundred MeV; there is no reliable observational indication on the energies of accelerated electrons. It is also in this phase that differential ionic acceleration, like that of He^3 ions can take place.

In the third phase particles are accelerated, probably by a second-order Fermi type process, in an open field line system that allows for the escape of accelerated particles to extra-solar space. This phase is obviously mainly related to eruptive/dynamic flares. The accelerated ions can reach GeV energies. This third phase occurs in a period of time ranging from one to several minutes after the first phase.

8.3. Status of Simulation Studies on Loops and Two-Loops Interaction

As summarized in Section 6, the *Yohkoh* observations yielded several evidences of two-loops interaction in solar flares. Among them, the flares of 9 December 1992 and 17 August 1992 are examples of I-type current loop coalescence. The flare of 2 August 1993 is an example of partial X-type coalescence. The flare of 9 December 1992 seems to be due to coalescence of two current-loops that do not carry equal currents. The observations, as well as the MHD simulations, lead consistently to the conclusion that the plasma density in a weak current-loop is enhanced during the approach of two loops. This is an example of energy being transmitted from one loop to another.

Present-days computer power is still insufficient for simulating fully realistic three-dimensional interactions of current-loops (curved loops with footpoints), based on a 3-D MHD code (Van Hoven, 1995) as well as on a 3-D particle code. But in the near future we may be able to study the interactions of curved loops, including a realistic treatment of their footpoints. It will then be important to strive for an internally consistent simulation of both phases: the generation phase of current-loops due to photospheric plasma motions (Parker, 1994; Sakai *et al.*, 1995) and the interaction phase of current-loops. It will also be important to simulate the nonlinear dynamics of a single curved current-loop including the footpoints (Hasegawa and Sakai, 1995), in order to study the mechanisms of energy release in a single loop.

As far as the study of processes of magnetic reconnection concerns, we need to clarify the processes of turbulent reconnection, including small-scale reconnection associated with the Hall effect and collision-less reconnection associated with micro-instabilities, especially in order to find the effective acceleration mechanisms of ions during solar flares.

It will also be worth while to simulate the interaction of realistic magnetic flux-loops in which the currents are not field-alligned. At the location of interaction and reconnection electric currents will originate (Somov, 1992) flowing mainly along magnetic field-lines, involving a separator. In general terms it seems timely to investigate the relative role of electric currents and their evolution, in relation to the evolution of the magnetic field structure of the active region.

Acknowledgements

We are particularly grateful to Drs M. Aschwanden and Z. Švestka for many helpful and constructive remarks. We also express our sincere thanks for help in preparing this paper to Drs. R.C. Canfield, E. L. Chupp, S. Enome, F. Farnik, T. Fushiki, T. Neubert, H. Nieuwenhuijzen, K. I. Nishikawa, M. Inda-Koide, S. Koide, K. Shibata, G. M. Simnett, B. V. Somov, M. Takahashi, and J. Zhao.

J.-I. Sakai acknowledges the partial support from the Cosel and Densoku Company and a Grant-in-Aid for scientific research (07640352) from the Ministry of Education of Japan. C. de Jager acknowledges support from the Japan Society for the Promotion of Science.

References

Acton, L. W., Feldman, U., Bruner, M. E., Doschek, G. A., Hirayama, T., Hudson, H. S., Lemen, J. R., Ogawara, T., Strong, K. T., and Tsuneta, S.: 1992, *Publ. Astron. Soc. Japan* **44**, L71.

Airapetian, V. S. and Smartt, R. N.: 1995, *Astrophys. J.* **445** 489.

Akinov, V. V., Belov, A. V., Chertok, I. M., Kurt, V. G., Leikov, N. G., Magun, A., and Melnikov, V. F.: 1993, *Proc. 23rd Int. Cosmic Ray Conf.* **3**, 111.

Alexander, D., Dunphey, Ph. P., and MAcKinnon, A. L.: 1994, *Solar Phys.* **151**, 147.

Antonucci, E., Dodero, M. A., and Martin, R.: 1990, *Astrophys. J. Suppl. Ser.* **73**, 137.

Antonucci, E., Dennis, B. R., Gabriel, A. H., and Simnett, G. M.: 1985, *Solar Phys.* **96**, 129.

Antonucci, E., Gabriel, A. H., and Dennis, B. R.: 1984, *Astrophys. J.* **287**, 917.

Aschwanden, M. J.: 1987, *Solar Phys.* **111**, 113.

Aschwanden, M. J. and Benz, A. O.: 1995, *Astrophys. J.* **438**, 997.

Aschwanden, M. J., Benz, A. O., Dennis, B. R., and Kundu, M.: 1992, *Astrophys. J.* **385**, 363.

Aschwanden, M. J., Benz, A. O., Dennis, B. R., and Gaizauskas, V.: 1993a, *Astrophys. J.* **416**, 857.

Aschwanden, M. J., Benz, A. O., and Schwartz, R. A.: 1993b, *Astrophys. J.* **417**, 790.

Aschwanden, M. J., Benz, A. O., and Montello, M. L.: 1994a, *Astrophys. J.* **431**, 432.

Aschwanden, M. J., Benz, A. O., Dennis, B. R., and Kundu, M. R.: 1994b, *Astrophys. J. Suppl. Ser.* **90**, 631.

Aschwanden, M. J., Hudson, H., and Schwartz, R. A.; 1995a, *Astrophys. J.*, submitted.

Aschwanden, M. J., Montello, M. L., Dennis, B. R., and Benz, A. O.: 1995b, *Astrophys. J.* **440**, 394.

Aschwanden, M. J., Schwarz, R. A., and Alt, D. M.: 1995c, *Astrophys. J.* **447**, 923.

Bagalá, L. G., Mandrini, C. H., Rovira, M. G., Démoulin, P., and Hénoux, J. C.: 1995, *Solar Phys.* **161**, 103.

Bai, T. and Sturrock, P. A.: 1989, *Ann. Rev. Astron. Astrophys.* **27**, 421.

Bai, T., Kiplinger, A. L., and Dennis, B. R.: 1985, *Bull. Amer. Astron. Soc.* **17**, 519.

Battacharje, A., Brunel, F., and Tajima, T.: 1983, *Phys. Fluids* **26**, 3332.

Beaufume, P., Coppi, B., and Golub, L.: 1992, *Astrophys. J.* **393**, 396.

Belov, A. V., Eroshenko, E. A., and Livshts, M. A.: 1996, *Proceedings Solar-Terr. Physics Symposium*, Sendai, Japan, in press.

Benka, S. and Holman, G. D.: 1992, *Astrophys. J.* **391**, 854.

Benz, A. O., Barrow, C. H., Pick, M., Raoult, A., and Simnett, G. M.: 1983, *Solar Phys.* **83**, 267.

Biskamp, D.: 1986, *Phys. Fluids* **29**, 1520.

Biskamp, D.: 1993, *Nonlinear Magnetohydrodynamics*, Cambrige University Press, Cambridge.

Biskamp, D.: 1994, *Phys. Reports*, No. 4, **237**, 179.

Biskamp, D. and Welter, H.: 1989, *Solar Phys.* **120**, 49.

Borman, P. and Lemen, J. R.: 1992, *Astrophys. J.* **385**, 363.

Bray, R. L., Cram, L. E., Durrant, C. J., and Loughhead, R. E.: 1991, *Plasma Loops in the Solar Corona*, Cambridge University Press, Cambridge, UK.

Brown, D. G. and Emslie, A. G.: 1987, *Solar Phys.* **96**, 143.

Brunel, F., Tajima, T., and Dawson, J. M.: 1982, *Phys. Rev. Letters* **49**, 323.

Bulanov, S. V., Sakai, J. I., and Syrovatskii, S. I.: 1979, *Soviet J. Plasma Phys.* **5**, 157.

Bumba, V. and Křivský, L.: 1959, *Bull. Astron. Inst. Czech.* **10**, 221.

Buneman, O.: 1993, in H. Matsumoto and Y. Omura (eds.), *Computer Space Plasma Physics, Simulation Techniques and Softwares*, Terra Scientific, Tokyo, p. 67.

Buneman, O., Neubert, T., and Nishikawa, K. I.: 1992, *IEEE Tran. Plasma Sci.* **20**, 810.

Burkepile, J. T. and St. Cyr, O. C.: 1993, *A Revised and Expanded Catalogue of Mass Ejections Observed by the Solar Maximum Mission Coronagraph*, N.C.A.R., Boulder.

Cane, H. V., Reames, D. V., and Von Rosenvinge, T.T.: 1988, *J. Geophys. Res.* **93**, 9555.

Canfield, R. C., Cheng, C. C., Dere, K. P., Dulk, G. A., McLean, D. J., Robinson, R. D., Schmall, E. J., and Schoolman, S. A.: 1980, in P. A. Sturrock (ed.), *Solar Flares*, Colorado Ass. University Press, Boulder, p. 451.

Canfield, R. C., Metcalf, T. R., Strong, K. T., and Zarro, D. M.: 1987, *Nature* **326**, 165.

Canfield, R. C., Hudson, H. S., Leska, K. D., Mickey, D. I., Metcalf, T. R., Wuesler, J. P., Acton, L. W., Strong, K. T., Kosugi, T., Sakao, T., Tsuneta, S., Culhane, J. L., Phillips, A., and Fludra, A.: 1992, *Publ. Astron. Soc. Japan* **44**, L11.

Chargeishvili, B., Zhao, J., and Sakai, J. I.: 1993, *Solar Phys.* **145**, 297.

Chen, J., Wang, H., Zirin, H., and Ai, G.: 1994, *Solar Phys.* **154**, 261.

Cheng, C. C. and Acton, L.: 1994, in Y. Uchida, T. Watanabe, K. Shibata, and H. S. Hudson (eds.), *X-ray Solar Physics from Yohkoh*, Universal Academic Press, Tokyo, Japan, p. 89.

Chupp, E. L.: 1983a, *Solar Phys.* **86**, 383.

Chupp, E. L.: 1983b, *Solar Phys.* **86**, 435.

Chupp, E. L.: 1984, *Ann. Rev. Astron. Astrophys.* **22**, 359.

Chupp, E. L.: 1990, *Astrophys. J. Suppl. Series* **73**, 213.

Chupp, E. L., Troffet, G., Marschhäuser, H., Pick, M., Soru-Escaut, J., Rieger, E., and Dunphy, P. P.: 1993, *Astron. Astrophys.* **275**, 602.

Cliver, E. W.: 1995, *Solar Phys.* **157**, 285.

Cliver, E. W., Kahler, S. W., Shea, M. A., and Smart, D. F.: 1982, *Astrophys. J.* **260**, 362.

Cliver, E. W., Dennis, B. R., Kiplinger, A., Kane, S., Neidig, D., Sheely, N., and Koomen, N.: 1986, *Astrophys. J.* **305**, 920.

Cliver, E. W., Crosby, N. B., and Dennis, B. R.: 1994, *Astrophys. J.* **426**, 767.

Correia, E., Costa, J. E. R., Kaufmann, P., Magun, A., and Herrmann, R.: 1995, *Solar Phys.* **159**, 143.

Culhane, J. L., Fludra, A., Bentley, R. D., Doschek, G. A., Watanabe, T., Hiei, Lang, J., Carter, M. K., Marosla, J. T., Phillips, A. T., Pike, C. D., and Sterling, A. C.: 1992, *Publ. Astron. Soc. Japan* **44**, L101.

de Jager, C.: 1969, in Z. Švestka and C. de Jager (eds.), *Solar Flares and Space Reserch*, North Holland Publ., Amsterdam, p. 1.

de Jager, C.: 1985a, *Solar Phys.* **96**, 143.

de Jager, C.: 1985b, *Solar Phys.* **98**, 267.

de Jager, C.: 1986a, *Space Sci. Rev.* **44**, 43.

de Jager, C.: 1986b, *Solar Phys.* **98**, 267.
de Jager, C.: 1987, *Solar Phys.* **110**, 337.
de Jager, C.: 1988, *Proc. 20th Cosmic Ray Conf., Moscow* **7**, 66.
de Jager, C.: 1990, *Adv. Space Res.* **10** (9), 101.
de Jager, C.: 1996, *J. Geophys. Japan*, in press.
de Jager, C. and Boelee, A.: 1984, *Solar Phys.* **92**, 227.
de Jager, C. and de Jonge, G.: 1978, *Solar Phys.* **58**, 127.
de Jager, C. and Kundu, M. R.: 1963, *Space Res.* **3**, 836.
de Jager, C. and Sakai, J. I.: 1991, *Solar Phys.* **133**, 395.
de Jager, C. and Švestka, Z.: 1985, *Solar Phys.* **100**, 435.
de Jager, C., Machado, M. E., Schadee, A., Strong, K. T., Švestka, Z., Woodgate, B. E., and Van Tend, W.: 1983, *Solar Phys.* **84**, 205.
de Jager, C., Boelee, A., and Rust, D. M.: 1984, *Solar Phys.* **92**, 245.
de Jager, C., Kuijpers, J., Correia, E., and Kaufmann, P.: 1987, *Solar Phys.* **110**, 337.
de Jager, C., Inda-Koide, M., Koide, S., and Sakai, J. I.: 1994, in J. I. Sakai (ed.), *Symposium on Current-Loop Interaction in Solar Flares*, Toyama University, p. 149.
de Jager, C., Inda-Koide, M., Koide, S., and Sakai, J. I.: 1995, *Solar Phys.* **158**, 391.
Den, O. E. and Kormenkov, G. I.: 1993, *Astron. Zh.* **70**, 141. (translated *Astrophys. Rep.* **37**, 76).
Dennis, B. R. and Zarro, D. C.: 1993, *Solar Phys.* **146**, 177.
Dennis, B. R., Holman, G. D., Hudson, H. S., Kosugi, T., Strong, K. T., and Zarro, D. M.: 1994, in S. Enome and T. Hirayama (eds.), *Proceedings of Kofu Symposium: New Look at the Sun*, Radioheliograph Series No. 6, Nobeyama Radio Obs., p. 217.
Doschek, C. A., Mariska, J. T., Watanabe, T., Hiei, E., Lang, J., Culhane, J. L., Bentley, R. D., Brown, C. M., Feldman, U., Phillips, A. T., Phillips, K. J. H., and Sterling, A. C.: 1992, *Publ. Astron. Soc. Japan* **44**, L95.
Dreschhoff, G. A. M. and Zeller, E. J.: 1990, *Solar Phys.* **127**, 333.
Dreschhoff, G. A. M., Zeller, E. J., Qin, D., and Parker, B. C.: 1993, *Adv. Space Res.* **13**, No. 9, 443.
Dryer, M.: 1994, *Space Sci. Rev.* **67**, 363.
Dungey, J. W.: 1961, *Phys. Rev. Letters* **6**, 47.
Duyveman, A., Hoyng, P., and Machado, M. E.: 1982, *Solar Phys.* **81**, 137.
Edwin, P. M.: 1991, *Ann. Géophys.* **9**, 188.
Ellison, M. A.: 1949, *Monthly Notices Roy. Astron. Soc.* **106**, 500.
Enome, S., Nakajima, H., Shibasaki, K., Nishio, M., Takano, T., Hanaoka, Y., Totii, C., Shiomi, Y., Sekiguchi, H., Bushimata, T., Kawashima, S., Shinohara, N., Irimajiri, Y., Koshiishi, H., Choi, Y. S., Sakai, J. I., Takahashi, M., Takakura, T., Sakao, T., and Kosugi, T.: 1994, *Publ. Astron. Soc. Japan*, **000**, L27.
Fang, C., Hénoux, J. C., Hu, J., Yin-Zhang, X., Gao, X., and Fu, Q.: 1995, *Solar Phys.* **157**, 271.
Farnik, F., Hudson, H., and Watanabe, T.: 1996, *Solar Phys.* **165**, in press.
Feynman, J. and Martin, S. F.: 1994, personal communication.
Foote, B. J. and Craig, I. J. D.: 1990, *Astrophys. J.* **350**, 437.
Forbes, T. G. and Acton, L. W.: 1996, *Astrophys. J.*, in press.
Forbes, T. G. and Isenberg, P. A.: 1991, *Astrophys. J.* **373**, 294.
Forbes, T. G. and Priest, E. R.: 1984, *Solar Phys.* **94**, 315.
Forrest, D. J. and and Chupp, T. A.: 1983, *Nature* **305**, 291.
Forrest, D. J., Vestrand, W. T., Chupp, E. L., Riger, E., Cooper, J., and Share, G. H.: 1986, *Adv. Space Res.* **6** (6), 115.
Friedman, H., Chubb, T. A., Kupperian, J. E., Kreplin, R. W., and Lindsay, J. C.: 1958, *IGY Rocket Report*, Ser. 1, 183.
Fushiki, T. and Sakai, J. I.: 1994, *Solar Phys.* **150**, 117.
Fushiki, T. and Sakai, J. I.: 1995a, *Solar Phys.* **156**, 265.
Fushiki, T. and Sakai, J. I.: 1995b, *Solar Phys.* **161**, 317.
Fushiki, T. and Sakai, J. I.: 1995c, PAFS-27 'Research Report on Plasma Astrophysics and Fusion Science', Toyama University, preprint
Gaizauskas, V. and Harvey, K. L.: 1986, *Adv. Space Res.* **6**, 17.
Gary, G. A. and Démoulin, P.: 1995, *Astrophys. J.* **445**, 982.

Gesztelyi, L. and Kalman, B.: 1986, *Adv. Space Res.* **6**, 21.

Giovanelli, R. G.: 1946, *Nature* **158**, 81.

Gold, T. and Hoyle, F.: 1960, *Monthly Notices Roy. Astron. Soc.* **120**, 89.

Gosling, J. T.: 1993, *J. Geophys. Res.* **98**, 18937.

Gosling, J. T. and Hundhausen, A. J.: 1995, *Solar Phys.* **160**, 57.

Goldman, M. V.: 1984, *Solar Phys.* **89**, 403.

Golub, L., Herant, M., Kalata, K., Lovas, I., Nystrom, G., Pardo, F., Spiller, E., and Wilczynski, J.: 1990, *Nature* **344**, 842.

Hachenberg, O.: 1958, *Z. Astrophys.* **46**, 67.

Hagyard, M. J.: 1992, personal communication.

Hanaoka, Y.: 1994, *Astrophys. J.* **420**, L37.

Hanaoka, Y., Kurokawa, H., Enome, S., Nakajima, H., Shibasaki, K., Nishino, M., Takano, T., Torii, C., Sekiguchi, H., Kawashima, S., Bushimata, T., Shinohara, N., Irimajiri, Y., Koshiishi, H., Shiomi, Y., Nakai, Y., Funakoshi, Y., Kitai, R., Ishiura, K., and Kimura, G.: 1994, *Publ. Astron. Soc. Japan* **46**, 205.

Hara, H., Tsuneta, S., Lemen, J. R., Acton, L. W., and McTiernan, J. M.: 1992, *Publ. Astron. Soc. Japan* **44**, L135.

Harrison, R. A.: 1991, *Adv. Space Res.* **11**, 25.

Harrison, R. A.: 1992, *Phil. Trans. Roy. Soc. London* **A336**, 401.

Harrison, R. A. and Sime, D. G.: 1992, in J. J. Hunt and R. Reinhardt (eds.), *Proc. 26th Eslab Symposium*, ESA SP-346, 289.

Harrison, R. A., Waggett, P. W., Bruner, M., Dryer, M., and Simnett, G. M.: 1985, *Solar Phys.* **97**, 387.

Hasegawa, Y. and Sakai, J. I.: 1995, *Research Report on Plasma Astrophysics and Fusion Science*, Toyama University.

Heyvaerts, J., Priest, E. R., and Rust, D. M.: 1977, *Astrophys. J.* **216**, 123.

Hirayama, T.: 1974, *Solar Phys.* **34**, 323.

Holman, G. D.: 1995, *Astrophys. J.* **452**, 451.

Holman, G. D., Eichler, D., and Kundu, M. R.: 1980, in M. R. Kundu and T. E. Gergely (eds.), 'Radio Physics of the Sun', *IAU Symp.* **86**, 457.

Hood, A. W. and Priest, E. R.: 1981, *Geophys. Astrophys. Fluid Dynamics* **17**, 297.

Howard, R. and Švestka, Z.: 1977, *Solar Phys.* **54**, 65.

Hoyng, P., Brown, J. C., and Van Beek, H. F.: 1976, *Solar Phys.* **48**, 197.

Hoyng, P., Duyveman, A., Machado, M. E., Rust, D. M., Švestka, Z., Boelee, A., de Jager, C., Frost, K. J., Lafleur, H., Simnett, G., van Beek, H. F., and Woodgate, B. E.: 1981, *Astrophys. J.* **246**, L115.

Hudson, H. S., Acton, L. W., Hirayama, T., and Uchida, Y.: 1992, *Publ. Astron. Soc. Japan* **44**, L77.

Ichimoto, K. and Kurokawa, H.: 1984, *Solar Phys.* **93**, 105.

Ichimoto, K., Hirayama, T., Yamaguchi, A., Kumagai, K., Tsuneta, S., Hara, H., Acton, L. W., and Bruner, M. E.: 1992, *Publ. Astron. Soc. Japan* **44**, L117.

Inda-Koide, M.: 1994, in J. I. Sakai (ed.): *Symposium on Current-Loop Interaction in Solar Flares*, Toyama University, p. 115.

Inda-Koide, M., Sakai, J. I., Koide, S. J., Kosugi, T., Sakao, T., and Shimizu, T.: 1995, *Publ. Astron. Soc. Japan* **47**, 323.

Kahler, S. W.: 1993, *J. Geophys. Res.* **98**, 5607.

Kahler, S.W.: 1994, *Astrophys. J.* **428**, 837.

Kahler, S. W., Moore, R. L., Kane, S. R., and Zirin, H.: 1988, *Astrophys. J.* **328**, 824.

Kahler, S. W., Daibog, E. J., Kurt, V. G., and Stopovskii, V. G.: 1994, *Astrophys. J.* **422**, 394.

Kai, K., Nakajima, H., and Kosugi, T.: 1983, *Publ. Astron. Soc. Japan* **35**, 285.

Kallenrode, M. B., Wibberenz, G., Kunov, H., Muller-Mellin, R., Stolpovskii, V., and Kontor, N: 1993, *Solar Phys.* **147**, 377.

Kallenrode, M. B. and Švestka, Z.: 1994, *Solar Phys.* **155**, 121.

Kanbach, G., Bertsch, D. L., and Fichtel, C. E.: 1993, *Astron. Astrophys. Suppl.* **97**, 349.

Kane, S. R., Kai, K., Kosugi, T., Enome, S., Landecker, P. B., and McKenzie, D. L.: 1983, *Astrophys. J.* **271**, 376.

Kane, S. R., Chupp, E. L., Forrest, D. J., Share, G. H., and Rieger, E.: 1986, *Astrophys. J.* **300**, L95.
Kano, R. and Tsuneta, S.: 1996, *Publ. Astron. Soc. Japan*, submitted.
Karlický, M. and Zhou A. H.: 1993, *Solar Phys.* **146**, 199.
Kawabata, K., Yoshimori, M., Suga, K., Morimoto. K., Hiraoka, T., Sato, J., and Ohki, K.: 1994, *Astrophys. J. Suppl. Ser.* **90**, 701.
Kiplinger, A. L., Dennis, B. R., Emslie, A. G., Frost, K. J., and Orwing, L. E.: 1983, *Astrophys. J.* **265**, L99.
Klimchuk, J. A., Lemen, J. R., Feldman, U., Tsuneta, S., and Uchida, Y.: 1992, *Publ. Astron. Soc. Japan* **44**, L181.
Kocharov, L. G. and Kocharov, G. E.: 1984, *Space Sci. Rev.* **38**, 89.
Kocharov, L. G. and Kovaltsov, G. A.: 1993, in G. P. Zank and T. K. Gaisser (eds.), *Particle Acceleration in Cosmical Plasmas*, A.I.P. Conf. Proc. 264, 239.
Kocharov, L. G., Lee, J. W., Zirin, H., Kovaltsov, G. A., Usoskion, I. G., Pyle, K. R., Shea, M. A., and Smart, D. F.: 1994, *Solar Phys.* **150**, 267.
Koide, S.: 1990, *J. Phys. Soc. Japan* **59**, 3952.
Koide, S. and Nishikawa, K. I.: 1995, PAFS-36; Res. Rept. Plasma Astrophys. Fusion Sci., Toyama University.
Koide, S. and Sakai, J. I.: 1994, *Solar Phys.* **151**, 137.
Koide, S., Sakai, J. I., and Chargeishvili, B.: 1994, in J. I. Sakai (ed.), *Symposium on Current-Loop Interaction in Solar Flares*, Toyama University, Japan, p. 55.
Kopp, R. A. and Pneuman, G. W.: 1976, *Solar Phys.* **50**, 85.
Kosugi, T.: 1993, *Yohkoh HXT Databook*, Astron. Obs. Mitaka, Tokyo.
Kosugi, T.: 1994, *Japanese J. Appl. Phys.* **33** (1), 2862.
Kovacz, A. and Dezso, L.: 1986, *Adv. Space Res.* **6**, 29.
Kuperus, M., Ionson, J. A., and Spicer, D. S.: 1981, *Ann. Rev. Astron. Astrophys.* **19**, 7.
Kuperus, M. and Van Tend, W.: 1981, *Solar Phys.* **71**, 125.
Kusano, K., Suzuki, Y., and Nishikawa, K.: 1994, INSAM report No.6, Hiroshima University, preprint.
Lau, Y. T. and Ramaty, R.; 1995, *Solar Phys.* **160**, 343.
LaRosa, T. N., Moore, R. E., and Shore, S. N.: 1994, *Astrophys. J.* **425**, 856.
LaRosa, T. N., Moore, R. E., Miller, J. A., and Shore, S. N.: 1996, *Astrophys. J.*, submitted.
Lee, J. W. and Gary, D. E.: 1994, *Solar Phys.* **153**, 347.
Leka, K. D., Canfield, R. C., McClymont, A. N., de la Beaujardière, J. F., Fan, W., and Tang, F.: 1993, *Astrophys. J.* **411**, 370.
Lemmens, A. and de Jager, C.: 1986, *Solar Phys.* **106**, 365.
Lim, J., Gary, D. E., Hurford, G. J., and Lemen, J. R.: 1993, *Bull. Amer. Astron. Soc.* **25**, 1198.
Lin, R. P. and Hudson, H. S.: 1971, *Solar Phys.* **17**, 412.
Lindman, E. L.: 1975, *J. Comp. Phys.* **18**, 66.
Litvinenko, Yu. E. and Somov, B. V.: 1995, *Solar Phys.* **158**, 317.
Loran, J. M. and Brown, J. C.: 1985, *Astrophys. Space Sci.* **117**, 173.
Low, B. C. and Wolfson, R.: 1988, *Astrophys. J.* **324**, 574.
Machado, M. E.: 1982, *Adv. Space Rev.* **2**, 115.
Machado, M. E., Duijveman, A., and Dennis, B. R.: 1982, *Solar Phys.* **79**, 85.
Mandzhavidze, N. and Ramaty, R.: 1993, *Nucl. Phys. B. Proc. Symp.* **33**, 141.
Masuda, S., Kosugi, T., Hara, H., Tsuneta, S., and Ogawara, Y.: 1994, *Nature* **371**, 495.
Martin, S. F. and Ramsey, H. E.: 1972, in P. S. McIntosh and M. Dryer (eds.), *Solar Activity Observations and Predictions*, M.I.T. Press, Cambridge, p. 371.
Matsumoto, R. and Shibata, K.: 1992, *Publ. Astron. Soc. Japan* **44**, 167.
Matsumoto, R., Tajima, T., Shibata, K., and Kaisig, M.: 1993, *Astrophys. J.* **414**, 357.
Meerson, B. I. and Sosorov, P. V.: 1981, *Adv. Space Res.* **1** (3), 77.
Matsusita, K., Masuda, S., Kosugi, T., Inda, M., and Yaji, K.: 1992, *Publ. Astron. Soc. Japan* **44**, L89.
McCabe, M. K.: 1985, *Solar Phys.* **98**, 127.
Melrose, D. B.: 1991, *Astrophys. J.* **381**, 306.
Mikhailovskii, A. B.: 1975, *Rev. Plasma Phys.* **6**, 77.
Mikic, Z., Barnes, C. D., and Schnack, D. D.: 1988, *Astrophys. J.* **328**, 830.
Mikic, Z. and Linker, J. A.: 1994, *Astrophys. J.* **430**, 898.

Mikic, Z., Schnack, D. D., and Van Hoven, G.: 1990, *Astrophys. J.* **361**, 690.
Miller, J. A. and Ramaty, R.: 1993, in G. P. Zank and T. K. Gaisser (eds.), *Particle Acceleration in Cosmic Plasmas*, A.I.P. Conf. Proc. 264, 223.
Miller, J. A., Guessoum, N., and Ramaty, R.: 1990, *Astropys. J.* **361**, 701.
Mouradian, Z., Soru-Escaut, I., and Pojoga, S.: 1995, *Solar Phys.* **158**, 269.
Murphy, R. J., Ramaty, R., Kozlovsky, B., and Reames, D. V.: 1991, *Astrophys. J.* **371**, 793.
Nakajima, H., Kosugi, T., Kai, K., and Enome, S.: 1983, *Nature* **305**, 292.
Nakajima, H., Dennis, B. R., Hoyng, P., Nelson, G., Kosugi, T., and Kai, K.: 1985, *Astrophys. J.* **288**, 806.
Nakajima, H.: 1994, in J. I. Sakai (ed.), *Symposium on Current-Loop Interaction in Solar Flares*, Toyama University, p. 39.
Nakariakov, V. M. and Roberts, B.: 1995, *Solar Phys.* **159**, 399.
Narain, U. and Ulmschneider, P.: 1990, *Space Sci. Rev.* **54**, 377.
Narain, U. and Ulmschneider, P.: 1996, *Space Sci. Rev.*, in press.
Neiding, D. F., Wiborg, P. H., and Gilliam, R. C.: 1993, *Astrophys. J.* **384**, 341.
Neubert, T., Miller, R. H., Buneman, O., and Nishikawa, K. I.: 1992, *J. Geophys. Res.* **97** (A8), 12057.
Neupert, W.: 1968, *Astrophys. J.* **153**, L59.
Nishikawa, K. I., Buneman, O., and Neubert, T.: 1993, private communication.
Nishikawa, K. I., Sakai, J. I., Zhao, J., Neubert, T., and Buneman, O.: 1994, *Astrophys. J.* **434**, 363.
Nishino, M., Nakajima, H., Enome, S., Shibasaki, K., Takano, T., Hanaoka, Y., Torii, C., Sekiguchi, H., Bushimata, T., Kawashima, S., Shinohara, N., Irumajiri, Y., Choi, Y., Koshiishi, H., Shiomi, Y., Metcalf, T. R., and Canfield, R. C.: 1994, *Publ. Astron. Soc. Japan* **46**, L11.
Nozawa, S., Shibata, K., Matsumoto, R., Sterling, A. C., Tajima, T., Uchida, Y., Ferrari, A., and Rosner, R.: 1992, *Astrophys. J. Suppl.* **78**, 267.
Ogawara, A., Acton, L. W., Bentley, D., Brumer, M. E., Culhane J. L., Hiei, E., Hirayama, T., Hudson, H. S., Kosug, T., Lemen, J. R., Strong, K. T., Tsuneta, S., Uchida, Y., Watanabe, T., and Yoshimori, M.: 1992, *Publ. Astron. Soc. Japan* **44**, L41.
Ohsawa, Y.: 1985, *Phys. Fluids* **28**, 2130.
Ohsawa, Y.: 1993, *J. Phys. Soc. Japan* **62**, 2382.
Ohsawa, Y. and Sakai, J. I.: 1987, *Astrophys. J.* **313**, 440.
Ohsawa, Y. and Sakai, J. I.: 1988a, *Solar Phys.* **116**, 157.
Ohsawa, Y. and Sakai, J. I.: 1988b, *Astrophys. J.* **332**, 439.
Ohsawa, Y. and Terashima, Y.: 1993, *Publ. Astron. Soc. Japan* **42**, 551.
Omura, Y. and Matsumoto, H.: 1988, *Geophys. Res. Letters* **15**, 319.
Ono, Y., Morita, A., and Katsurai, M.: 1993, *Phys. Fluids* **B5**, 3691.
Pallavicini, R., Serio, S., and Vaiana, G. S.: 1977, *Astrophys. J.* **216**, 108.
Parker, E. N.: 1963, *Astrophys. J. Suppl. Ser.* **8**, 177.
Parker, E. N.: 1988, *Astrophys. J.* **330**, 474.
Parker, E. N.: 1994, *Spontaneous Current-Sheets in Magnetic Fields, with Applications to Stellar X-rays*, Oxford, UK.
Piana, M., Brown, J. C., and Thompson, A. M.: 1995, *Solar Phys.* **156**, 315.
Pérez Enriquez, R. and Mendoza, B.: 1995, *Solar Phys.* **160**, 353.
Petschek, H. E.: 1964, in W. H. Ness (ed.), *AAS/NASA Symp. on the Physics of Solar Flares*, NASA, Washington, DC, p. 425.
Pesses, M. E., Klecker, B., Glocker, G., and Hovestadt, D.: 1981, *17th Int. Cosmic Ray Conf. Paris* **3**, 36.
Phan, T. D. and Sonnerup, B. U. O.: 1991, *J. Plasma Phys.* **46**, 407.
Porter, J. G., Fontenla, J. M., and Simnett, G. M.: 1995, *Astrophys. J.* **438**, 472.
Priest, E. R.: 1985, *Rep. Prog. Phys.* **48**, 955.
Pritchett, P. L.: 1990, *Phys. Fluids* **B4** (10), 3371.
Pritchett, P. L., Karamabadi, H., and Omidi, N.: 1989, *Geophys. Res. Letters* **16**, 883.
Ramaty, R.: 1993, *Adv. Space Res.* **13**, 275.
Ramaty, R. and Murphy, R. J.: 1987, *Space Sci. Rev.* **45**, 213.
Ramaty, R., Schwartz, R. A., Enome, S., and Nakajima, H.: 1995, *Astrophys. J.* **436**, 941.

Reames, D. V.: 1993, in G. P. Zank and T. K. Gaisser (eds.), *Particle Acceleration in Cosmic Plasmas*, A.I.P. Conf. Proc. 264, 213.
Rieger, E: 1989, *Solar Phys.* **121**, 323.
Rompolt, B.: 1984, *Adv. Space Res.* **4** (7), 357.
Rust, D. M. and Švestka, Z.: 1979, *Solar Phys.* **63**, 279.
Sakai, J. I.: 1982, *Astrophys. J.* **263**, 970.
Sakai, J. I.: 1983a, *J. Plasma Phys.* **30**, 109.
Sakai, J. I.: 1983b, *Solar Phys.* **84**, 109.
Sakai, J. I.: 1989, *Solar Phys.* **20**, 117.
Sakai, J. I.: 1990a, *Astrophys. J.* **365**, 354.
Sakai, J. I.: 1990b, *Astrophys. J. Suppl.* **73**, 321.
Sakai, J. I.: 1992, *Solar Phys.* **140**, 99.
Sakai, J. I. and de Jager, C.: 1989a, *Solar Phys.* **123**, 393.
Sakai, J. I. and de Jager, C.: 1989b, *Solar Phys.* **123**, 389.
Sakai, J. I. and de Jager, C.: 1991, *Solar Phys.* **134**, 329.
Sakai, J. I. and de Jager, C.: 1996, *Solar Phys.* in preparation.
Sakai, J. I. and Fushiki, T.: 1995a, *Solar Phys.* **156**, 281.
Sakai, J. I. and Fushiki, T.: 1995b, PAFS-34, *Research Report on Plasma Astrophysics and Fusion Science*, Toyama University.
Sakai, J. I., Fushiki, T., and Nishikawa, K. I.: 1995, *Solar Phys.* **158**, 301.
Sakai, J. I. and Koide, S.: 1992, *Solar Phys.* **137**, 293.
Sakai, J. I. and Koide, S.: 1992b, *Solar Phys.* **142**, 399.
Sakai, J. I. and Nishikawa, K. I.: 1983, *Solar Phys.* **88**, 241.
Sakai, J. I. and Ohsawa, Y.: 1987, *Space Sci. Rev.* **46**, 113.
Sakai, J. I., Suzuki, M., and Fushiki, T.: 1995, *IAU Colloq.* **153**.
Sakai, J. I., Tajima, T., and Brunel, F.: 1984, *Solar Phys.* **91**, 103.
Sakai, J. I. and Washimi, H.: 1982, *Astrophys. J.* **258**, 823.
Sakai, J. I., Zhao, J., and Nishikawa, K. I.: 1994, *Solar Phys.* **154**, 97.
Sakao, T.: 1993, Ph.D. Thesis, University of Tokyo.
Sakao, T: 1994, in J. I. Sakai (ed.), *Symposium on Current-Loop Interaction in Solar Flares*, Toyama University, p. 107.
Sakao, T., Kosugi, T., Masuda,, S., Inda, M., Makishima, K., Canfield, R. C., Hudson, H. S., Metcalf, T. R., Wuelser, J. P., Acton, L. W., and Ogawara, Y.: 1992, *Publ. Astron. Soc. Japan* **44**, L83.
Sakurai, T., Shibata, K., Ichimoto, K., Tsuneta, S., and Acton, L. W.: 1992a, *Publ. Astron. Soc. Japan* **44**, L123.
Sakurai, T., Ichimoto, K., Hiei, E., Irie, M., Kumagai, K., Miyashita, M., Nishino, Y., Yamaguchi, K., Fang, G., Kambry, M. A., Zhao, J., and Shinoda, K.: 1992b, *Publ. Astron. Soc. Japan* **44**, L7.
Seely, J. F. and Feldman U.: 1992, *Astrophys. J.* **394**, 697.
Shea, M. A., Smart, D. F., Allen, J. H., and Wilkinson, D. C.: 1992, *IEEE Trans. Nucl. Sci.* **39**, 1754.
Sheeley, J. R., Bohlin, J. D., Bruckner, G. E., Purcell, J. D., Scherrer, V., and Tousey, R.: 1975, *Astrophys. J.* **196**, L129.
Shibata, K., Ishido, Y., Acton, L. W., Strong, K. T., Hirayama, T., Uchida, Y., McAllister, A. H., Matsumoto, R., Tsuneta, S., Shimizu, T., Hara, H., Sakurai, T., Ichimoto, K., Nishio, Y., and Ogawara, Y.: 1992, *Publ. Astron. Soc. Japan* **44**, L173.
Shibata, K., Masuda, S., Shimojo, M., Hara, H., Yokoyama, T., Tsuneta, S., Kosugi, T., and Ogawara, Y.: 1995, *Astrophys. J. Letters*, in press.
Shibata, K., Nitta, N., Strong, K. T., Matsumoto, R., Yokoyama, T., Hirayama, T., Hudson, H., and Ogawara, Y.: 1994a, *Astrophys. J.* **431**, L51.
Shibata, K., Yokoyama, T., and Shimojo, M.: 1994b, *Proc. of Kofu Symposium*, NRO Report No.360, p. 75.
Shibata, K. *et al.*: 1996, Paper at Sendai Solar Terr. Phys. Symposium, preprint.
Shimizu, T., Tsuneta, T., Acton, L. W., Lemen, J. R., and Uchida, Y.: 1992, *Publ. Astron. Soc. Japan* **44**, L147.
Shimizu, T., Tsuneta, T., Acton, L. W., Lemen, J. R., Ogawara, Y., and Uchida, Y.: 1994, *Astrophys. J.* **422**, 906.

Šimberová, S., Karlický, M., and Švestka, Z: 1993, *Solar Phys.* **146**, 343.
Simnett, G. M.: 1993, *Adv. Space Res.* **13**, No. 9, 133.
Simnett, G. M.:1995, *Space Sci. Rev.* **73**, 387.
Simnett, G. M. and Harrison, R. A.: 1985, *Solar Phys.* **99**, 291.
Simnett, G. M., Sotirovsky, P., and Simon, G.: 1990, *Astron. Astrophys.* **227**, 235.
Smart, D. F. and Shea, M. A.: 1989, *J. Spacecraft and Rockets* **26**, 403.
Smartt , R. N., Zang, Z., and Smutko, M. F.: 1993, *Solar Phys.* **148**, 139.
Somov, B. V.: 1992, *Physical Processes in Solar Flares*, Kluwer Academic Publishers, Dordrecht, Holland.
Somov, B. V. and Syrovatskii, S. I.: 1974, *Solar Phys.* **39**, 415.
Somov, B. V. and Syrovatskii, S. I.: 1977, *Solar Phys.* **39**, 393.
Somov, B. V. and Syrovatskii, S. I.: 1982, *Solar Phys.* **75**, 237.
Somov, B. V. and Titov, V. S.: 1985a, *Solar Phys.* **95**, 141.
Somov, B. V. and Titov, V. S.: 1985b, *Solar Phys.* **102**, 79.
Somov, B. V. and Verneta, A. I.: 1988, *Solar Phys.* **117**, 89.
Somov, B. V. and Verneta, A. I.: 1989, *Solar Phys.* **121**, 321.
Spicer, D. S.: 1976, *An Unstable Arch Model of a Solar Flare*, NRL Report 8036.
Spicer, D. S.: 1982, *Space Sci. Rev.* **31**, 351.
StCyr, O. C. and Webb, D. F.: 1991, *Astrophys. Space Sci.* **119**, 337.
Stoker, P. H.: 1995, *Space Sci. Rev.* **73**, 327.
Strauss, H. R.: 1990, *J. Geophys. Res.* **95**, 145.
Sturrock, P. A. (ed.): 1986, *Solar Flares: Monograph from Skylab Solar Workshop II*, Colorado Ass. University Press.
Sturrock, P. A.: 1989, *Solar Phys.* **121**, 387.
Sturrock, P. A., Holzer, T. E., Mihalas, D. M., and Ulrich, R. K. (eds.): 1980, *Physics of the Sun*, Vols. 1, 2, 3, D. Reidel Publ. Co., Dordrecht, Holland;
Suemoto, Z. and Hiei, E.: 1959, *Publ. Astron. Soc. Japan* **11**, 185.
Švestka, Z.: 1951, *Bull. Astron. Inst. Czech.* **2**, 100 and 120.
Švestka, Z.: 1976, *Solar Flares*, D. Reidel Publ. Co., Dordrecht, Holland.
Švestka, Z.: 1984, *Solar Phys.* **94**, 171.
Švestka, Z.: 1985, in D. Neiding (ed.), *Proceedings of the NSO/SMM Summer Meeting on the Low Temperature Flare Plasma*, Sacramento Peak.
Švestka, Z.: 1989, *Solar Phys.* **121**, 399.
Švestka, Z.: 1995, *Solar Phys.* **160**, 53.
Švestka, Z. and Howard, R.: 1981, *Solar Phys.* **71**, 349.
Švestka, Z., Stewart, R., Hoyng, P., van Tend, W., Acton, L. W., Gabriel, A. H., and Rapley, C. G.: 1982a, *Solar Phys.* **75**, 305.
Švestka, Z., Dennis, B. R., Pick, M., Raoult, A., Rapley, C. G., Stewart, R. T., and Woodgate, B. E.: 1982b, *Solar Phys.* **80**, 143.
Švestka, Z., Fontenla, J. M., Machado, M. E., Martin, S. F., Neidig, D. F., and Poletto, G.: 1987, *Solar Phys.* **108**, 237.
Švestka, Z., Farnik, F., Hudson, H., Uchida, Y., Hick, P., and Lemen, J. R.: 1995, *Solar Phys.* **161**, 331.
Sweet, P. A.: 1958, *Nuovo Cimento Suppl.* **8**, 188.
Syrovatsky, S. I.: 1966, *Soviet Astron.* **10**, 270.
Tajima, T. and Sakai, J. I.: 1989a, *Soviet J. Plasma Phys.* **15**, No. 8, 519.
Tajima, T. and Sakai, J. I.: 1989b, *Soviet J. Plasma Phys.* **15**, No. 9, 606.
Tajima, T., Brunel, F., and Sakai, J. I.: 1982, *Astrophys. J.* **245**, L45.
Tajima, T., Sakai, J. I., Nakajima, H., Kosugi, T., Brunel, F., and Kundu, M. R.: 1987, *Astrophys. J.* **321**, 1031.
Tajima, T., Benz, A. O., Thaker, M., and Leboeuf, J. N.:1990, *Astrophys. J.* **353**, 666.
Takahashi, M., Watanabe, T., Sakai, J. I., Tsuneta, S., Sakao, T., Kosugi, T., Sakurai, T., Enome, S., Nitta, N., Hudson, H., and Hashimoto, S.: 1995, *Publ. Astron. Soc. Japan*, submitted.

Takano, T., Enome, S., Nakajima, H., Shibasaki, K., Nishio, M., Hanaoka, Y., Torii, C., Sekiguchi, H., Bushimata, T., Kawashima, S., Shinohara, N., Irimajiri, Y., Koshiishi, H., Kosugi, T., Shiomi, Y., Sakurai, T., and Ishimoto, K.: 1994, *Publ. Astron. Soc. Japan* **46**, L21.

Takakura, T.: 1992, *Solar Phys.* **142**, 327.

Takakura, T., Inda, M., Makishima, K., Kosugi, T., Masuda, S., Sakurai, T., and Ogawara, Y.: 1993, *Publ. Astron. Soc. Japan* **45**, 737.

Takakura, T., Nishio, M., Nakajima,H., Enome, S., Shibasaki, K., Takano, T., Hanaoka, Y., Kosugi, T., Sakao, T., Masuda, T., Makishima, K., and Inda-Koide, M.: 1994, *Publ. Astron. Soc. Japan* **46**, 653.

Tanaka, K., Watanabe, T., Nishi, K., and Akita, K.: 1982, *Astrophys. J.* **254**, L59.

Tandberg-Hanssen, E. and Emslie, P. G.: 1988, *The Physics of Solar Flares*, Cambridge University Press, Cambrige, UK.

Tappin, S. J.: 1991, *Astron. Astrophys. Suppl.* **87**, 277.

Trottet, G., Chupp, E. L., Marschhaäuser, H., Pick, M., Soru-Escaut, I., Rieger, E., and Dunphy, P. P.: 1994, *Astron. Astrophys.* **288**, 647.

Tsuneta, S.: 1985, *Astrophys. J.* **290**, 353.

Tsuneta, S.: 1993, in H. Zirin, G. Ai, and H. Wong (eds.), *The Magnetic and Velocity Fields of Solar Active Regions*, Astron. Soc. Pacific, San Francisco. p. 239.

Tsuneta, S.: 1994, in Y. Uchida (ed.), *X-ray Solar Physics from Yohkoh*, Universal Academic Press, Tokyo, p. 115.

Tsuneta, S.: 1995, *Publ. Astron. Soc. Japan* **47**, 691.

Tsuneta, S., Nitta, N., Ohki, K., Takakura, T., Tanaka, K., Makashita, K., Murakami, T., Oda, M., and Ogawara, Y.: 1984, *Astrophys. J.* **284**, 827.

Tsuneta, S., Hara, H., Shimizu,,T., Acton, L. W., Strong, K. T., Hudson, H. S., and Ogawara, Y.: 1992a, *Publ. Astron. Soc. Japan* **44**, L63.

Tsuneta, S., Takahashi, T., Acton, L. W., Bruner, M., Harvey, K., and Ogawara, Y.: 1992b, *Publ. Astron. Soc. Japan* **44**, L211.

Tsuneta, S. and Lemen, J. R.: in J. F. Linsky and S. Serio (eds.), 1993, *Physics of Solar Flares and Stellar Coronae*, Kluwer Academic Publishers, Dordrecht, Holland, p. 113.

Uchida, Y.: 1996, Paper at Solar Terr. Phys. Symposium, Sendai, Japan, preprint.

Uchida, Y. and Shibata, K.: 1988, *Solar Phys.* **116**, 291.

Uchida, Y., McAllister, A., Strong, K. Y., Ogawara, Y., Shimizu, T., Matsumoto, R., and Hudson, H. S.: 1992, *Publ. Astron. Soc. Japan* **44**, L155.

Uchida, Y., Watanabe, T., and Yoshimori, M.: 1992, *Publ. Astron. Soc. Japan* **44**, L41.

van Beek, H. F., de Feiter, L. D., and de Jager, C.: 1974, *Space Research* **XIV**, 447

Van Hoven, G.:1995, *IAU Colloq.* **153** (paper).

van Tend, W. and Kuperus, M.: 1978, *Solar Phys.* **59**, 115.

Vekstein, G. E. and Priest, E.: 1992, *Astron. Astrophys.* **243**, 492.

Velli, M., Einaudi, G., and Hood, A. W.: 1990, *Astrophys. J.* **350**, 428.

Vorpahl, J. A.: 1976, *Astrophys. J.* **205**, 868.

Wagner, J. S., Lee, L. C., Wu, C. S., and Tajima, T.: 1983, *Geophys. Res. Letters* **10**(6), 483.

Wang, H.: 1992, *Solar Phys.* **140**, 85.

Wang, H., Gary, D. E., Lin, J., and Schwartz, R. A.: 1994, *Astrophys. J.* **433**, 379.

Watanabe, T., Kozuka, Y., Ohyama, M., Kojima, M., Yamachugi, K., Watari, S. T., Tsuneta, S., Joselyn, J. L., Harvey, K. L., Acton, L. W., and Klimchuk, J. A.: 1992, *Publ. Astron. Soc. Japan* **44**, L199.

Way, M.: 1994, personal communication.

Webb, D. F., Holman, G. D., Davis, J. M., Kundu, M. R., and Shevgaonkar, R. K.: 1987, *Astrophys. J.* **315**, 716.

Webb, D. F., Forbes, T. G., Aurass, H., Chen, J., Martens, P., Rompolt, B., Rušin, V., and Martin, S. F.: 1994, *Solar Phys.* **153**, 73.

Wild, J. P.: 1965, in W. N. Hess (ed.), *AAS/NASA Symposium on Physics of Solar Flares*, NASA, Washington, p. 103.

Wild, J. P. and McCready, L. L.: 1950, *Australian J. Sci. Res. Series* **A3**, 387.

Winglee, R. M., Kiplinger, A. L., Zarro, D. M., Dulk, G. A., and Lemen, J. R.: 1991, *Astrophys. J.* **375**, 306.

Winglee, R. M., Meninetti, J. D., and Wong, H. K.: 1992, *J. Geophys. Res.* **97**, 17131.

Wong, H. K. and Goldstein, M. L.: 1990, *Geophys. Res. Letters* **17**, 2229.

Wu, C. S. and Lee, L. C.: 1979, *Astrophys. J.* **230**, 612.

Wu, S. T., Weng, F. S., Wang, H. M., Zirin, H., and Ai, G. X.: 1993a, *Adv. Space Res.* **13** (9), 127.

Wu, S. T., Weng, F. S., Hagyard, M. J., Machado, M., and Schmieder, B.: 1993b, *Int. Astron. Union Colloq.* **144**, 7.

Wuelser, J. P., Zarro, D. M., and Canfield, R. C.: 1992, *Astrophys. J.* **384**, 341.

Yamada, M., Ono, Y., Hayakawa, A., Katsurai, M., and Perkins, F. W.: 1990, *Phys. Rev. Letters* **65**, 721.

Yokoyama, T. and Shibata, K.: 1994, *Proc. of Kofu Symposium*, NRO Report No. 360, p. 367.

Yokoyama, T. and Shibata, K.: 1995, *Nature* **375**, 42.

Yoshimori, M.: 1992, *Publ. Astron. Soc. Japan* **44**, L107.

Yoshimori, M.: 1994, in J. I. Sakai (ed.): *Symposium on Current-Loop Interaction in Solar Flares*, Toyama University, p. 45.

Yoshimori, M., Watanabe, H., and Takahashi, K.: 1986, *J. Phys. Soc. Japan* **55**, 471.

Yoshimori, M., Suga, K., Morimoto, K., Hiraoka, T., Sato, J., Kawabata, K., and Ohki, K.: 1994, *Astrophys. J. Suppl. Ser.* **90**, 639.

Yurchishin, V. B.: 1994, *Solar Phys.* **154**, 199.

Zaidman, E. G. and Tajima, T.: 1991, *Astrophys. J.* **338**, 1139.

Zaitsev, V. V. and Stepanov, A. V.: 1985, *Solar Phys.* **99**, 313.

Zarro, D. M., Mariska, J. T., and Dennis, D. R.: 1995, *Astrophys. J.* **440**, 888.

Zhang, T. X. and Ohsawa, Y.: 1995, *Solar Phys.* **158**, 115.

Zhao, J., Chargeishvili, B., and Sakai, J. I.: 1993a, *Solar Phys.* **146**, 331.

Zhao, J., Chargeishvili, B., and Sakai, J. I.: 1993b, *Solar Phys.* **147**, 131.

Zhao, J., Sakai, J. I., and Nishikawa, K. I.: 1995a, *Solar Phys.*, submitted.

Zhao, J., Sakai, J. I., and Nishikawa, K. I.: 1995b, *Solar Phys.*, submitted.

Zhao, J., Sakai, J. I., and Nishikawa, K. I.: 1995c, *Astrophys. J.* **449**, L161.

Zirin, H. and Tang, F.:1991, *Astrophys. J. Suppl.* **73**, 111.

Zirin, H. and Wang, H.: 1993, *Nature* **363**, 426.

Zwingmann, W.: 1987, *Solar Phys.* **111**, 309.

Made in United States
Troutdale, OR
11/13/2024

24743005R10115